Planets, Sunspots and Earthquakes

Planets, Sunspots and Earthquakes

Effects on the sun, the earth and its inhabitants

FRANK GLASBY

Includes 50 diagrams

Writers Club Press

San Jose   New York   Lincoln   Shanghai

Planets, Sunspots and Earthquakes
Effects on the sun, the earth and its inhabitants

Writers Club Press
an imprint of iUniverse, Inc.

For information address:
iUniverse, Inc.
5220 S. 16th St., Suite 200
Lincoln, NE 68512
www.iuniverse.com

ISBN: 0-595-22641-8

Printed in the United States of America

CONTENTS

LIST OF TABLES

PREFACE

The hypothesis of the influence of the planets on sunspots and earth-quakes considers two distinct and separate effects. One is an internal effect whereby the core of the sun and the earth become more active. The other is an external effect whereby the combined effect from external bodies stresses the earth's crust at a point where it enters or leaves the gravitational field of the group of external bodies. I wrote a paper on this, entitled *The Influence of Planetary Bodies on Earthtides and Earthquakes (1979)*. The table of the relative mass of the planets raised questions as to whether there could be a significant effect in terms of the conventional formula for external gravitational effects. However, further research showed that the standard formula is not applicable to this effect. An explanation of this is in the copy of the paper, together with two others on ocean tides and volcanoes, on the website of Darch Literary Service: <u>darchliterary.iinet.net.au</u>

In presenting these details, I wish to acknowledge that some of the conclusions presented in this work may not be correct. Students and researchers should check the details before applying any of the principles described here. This most particularly applies to any attempts to predict earthquakes. The author does not suggest that this knowledge will enable anyone to predict a specific earthquake. With formal scientific application, the principles should be helpful to specialists, and the work is set out with that purpose in mind, therefore writers should not quote the hypothesis as fact unless confirmed by further scientific research.

INTRODUCTION

Graphs of Seismic and Solar Activity

The first known attempt at foreseeing earthquakes was in China, in 132 AD. An investigator named Chang Heng devised a seismoscope for detecting seismic activity. This was a system where brass balls rolled out of their shallow cups when movement occurred. Yet, scientists only developed a more sophisticated device in the 19th century. Present day electronic equipment can now detect minute tremors in the earth's crust, and there are over a million such tremors every year. Only major earthquakes that affect human life make any news and overall there are only a few serious earthquakes each year. The main cause is from changes inside the earth, however there are some external causes, which are the basis of the following work. This is the result of research and observations over a period of three sunspot cycles, which showed a connection between solar and seismic activity and indicated the processes involved in that connection.

The investigation started with a project by my High School students, in 1964. From scientific records, we built two graphs on the same scale and compared earthquake and sunspot activity over a fifty-year period. These are in Figure 1. The graphs show smoothed curves that are almost the same. This coincidence suggested that there could be a connection between solar and seismic disturbances. However, books on solar and seismic phenomena gave no indication of any link and the texts treated solar and seismic activity as separate self-contained subjects. Further investigation suggested that a common effect might exist via the planets and with the possibility of a planetary connection; the search was for any matching relationships with solar and seismic events, then for indications of their significance. Despite some success in this, there was no firm connection in terms of physics until 1978. There were statistics but no scientific processes. Eventually, explanations that showed how the planets help to cause solar and seismic disturbance began to emerge. The first indication was of a subtle gravitational effect on which I published a paper but many exceptions indicated another essential process to which the gravitational effect is secondary. This eventually became clear enough to state as a principle. The explanations and examples of these points are in the following descriptions. They are set out as an aid to further study and scientific use in terms of practical application and possible prediction or forewarning of imminent solar and seismic events.

The Role of the Planets

The technical basis for the hypothesis of effects is very simple, although scientific proof is probably very complex. It rests on the premise that all the planets radiate some type of electromagnetic waves. The proposed explanation is that these electromagnetic forces affect the core of the sun and the core of the earth. The hypothesis is that the forces stimulate a radioactive process that leads to sunspots on the sun and earthquakes on the earth. Furthermore, gravitational forces from the planets trigger flares from the sunspots and earthquakes in stressed areas on the crust of the earth. This suggests that knowledge of how planets cause effects could aid in foreseeing solar and seismic activity, but specific prediction of seismic activity would depend on local scientific knowledge, as each fault behaves differently.

To start with, the question of possible biological effects did not arise, but references to animal disturbance before earthquakes eventually added a new dimension to the project. Scientists know of such effects, but not always as associated effects because they are generally under laboratory conditions relating to animal experiments. The discussion on these effects is therefore separate, and helps to show how the various effects interconnect. This interconnection helps to explain the various phenomena in this study. However, biological details are not the main emphasis; that is reserved for seismic and solar disturbances. To help to keep these ideas clear they are set out below in a more formal manner. The central hypotheses are as follows.

The Basic Principles of the Planetary Forces

- Planets emanate an electromagnetic radiation that, under certain specific conditions, stimulates the sun's core and help to cause sunspots.
- The core of the earth receives stimulation in a similar manner and this helps to create earthquakes, ie. the core becomes hotter, the magma becomes softer and the tectonic plates can then move more easily.
- In addition, specific combined planetary gravitational forces help to trigger flares on the surface of the sun by disturbing the sunspots.
- A similar effect occurs with the earth, which stress the seismic faults as they move in and out of the combined lunar, solar and planetary gravitational fields.
- The ocean tides add to the effects by increasing the horizontal compression caused by the specific gravitational effect.
- There are also biological effects from the combined electromagnetic forces.

Two Main Effects from the Planets

To sum up there are two basic principles. One is that electromagnetic radiation from the planets affects the core of the sun and the core of the earth. The other is that combined planetary gravitational forces disturb the surface of the sun and the surface of the earth. However this gravitational effect is not direct but is at a tangent to the line of direct traction. In earth tide effects, this is generally not at the high tide position but at the low tide position. In other words the planets, (and the sun and the moon for the earth) are mainly setting or rising on the affected area. The disturbed area is then either just rotating into, or rotating out of, the combined gravitational field. The principle is the same for the sun and for the earth but the differences are definable.

As the principles are closely similar for both the earth and the sun, the study of seismic activity helps to clarify the internal processes in the sun. Conversely, a study of the sun helps to clarify what happens inside the earth. In both cases, the likely effect is foreseeable although other factors can affect the outcome. The details are in the following chapters.

CHAPTER 1

SUNSPOT CYCLES AND THE PLANETS

Planetary Positions and Solar Activity

Almost everyone has heard of sunspots but few people really know what they are. Science books describe them as huge areas of turbulence that appear on the sun's surface about every eleven-years. We know that the disturbance starts within the sun and rises to the surface yet the cause is apparently still a mystery. The visible sunspots and their cyclic appearances are well known and the approximate eleven-year cycle is common scientific knowledge. However, a full sunspot cycle really takes twenty-two years because there are regular alternating changes in that period. One is in the magnetic field of the sun. For example in one eleven-year period, the solar North Pole is positive, but in the next, it is negative. Then it changes back. This makes the complete cycle of twenty-two years.

There is also a secondary magnetic effect. The sunspots appear in pairs in each hemisphere, ie. there are two pairs. One pair is in the northern half and the other pair is in the southern half. Each pair is also negative and positive. In one hemisphere there may be a positive spot leading and in the other hemisphere a negative spot leading. When the solar main field reverses its polarity, the next series of sunspots also reverse their order of polarity. Where positive was leading, negative now leads and conversely the same for the other hemisphere.

Four Key Times in Sunspot Activity

These changes do not occur at the same time. The main field tends to change nearer maximum activity whereas the sunspots change their polarity nearer to sunspot minimum activity. There is therefore the point of Sunspot Minimum activity, the point in time of Sunspot Maximum activity and two changes in polarity of the main field and of the sunspots. From these specific timing periods, the indication was that there should be repetitive planetary alignments. These should then coincide with solar changes. To start with, I studied planetary positions for the time of Sunspot Minimum and Sunspot Maximum in every cycle for the past two hundred years. The more detailed study was for a period of one hundred years because records were more easily accessible for the later period. For the eleven-year cycles, the planets Jupiter and Saturn emerged as the two main planets that made regular matching alignments.

On a longer time scale, Uranus and Neptune were the two main planets. The longer cycle was a period of approximately one hundred and eighty years. This operated in two half cycles of about 90 years in a similar manner to the way the eleven to twenty-two years cycles operate for the other two key planets; Jupiter and Saturn.

The Influence of Jupiter and Saturn

Of the two pairs of planets, Jupiter and Saturn were of special interest because they made suitable alignments more often. That is to say in every eleven-year cycle, as opposed to the greater 90-year cycle of the other pair of planets. Both Jupiter and Saturn have strong magnetic fields and they both emit a strong radio noise. The noise recorded by radio telescopes indicates that there is some type of electromagnetic radiation. Yet, studies of any effects from such radiation were hard to find. However, it seemed there might be an effect at the focus, where two radiation forces intersect. Could it be heat, nuclear stimulation, or both? This is a central question in this inquiry. The indication was that the focus of the intersection of radiation from two planets at right angles to each other might be a point of activity. Perhaps such an electromagnetic stimulation could cause a specific effect. We shall consider this later.

Nevertheless, the angular alignment of 90 degrees apart is only an ideal configuration. In practical operation, these two planets, Jupiter and Saturn, do not operate in isolation. There are other planets and the planets nearer to the sun than Jupiter appear to have an extra effect. This suggests a possible "timing mechanism" in which the faster moving planets have a function. From analysis, the two main planets were never in the ideal position and 90 degrees only appears as a general average.

Comparison of Alignments at Key Times

Figure 2a shows the relative positions of the Jupiter and Saturn at the times of maximum sunspot activity. This is for the period from 1870 to 1968. In the diagram, the short arm is Jupiter and the long arm is Saturn. The length of the arms indicates their relative distance from the sun. Jupiter is nearer. The angles are correct but for simplicity, the coordinate figures are not there. In addition, for the sake of comparison Saturn is in the same diagrammatic position at "six o'clock". They all relate to zero degrees on a heliocentric basis where zero is at the first point of the constellation of Aries, and the sun is at the centre. The planets are moving anti clockwise round the sun and Jupiter is the faster planet.

Four Basic Planetary Alignments

From this approach, there are four basic patterns. One is with the angle opening towards 180 degrees. Another is when the planets are closing to be in the same directional alignment. The other two are in the visual position whereby there is a left hand and a right hand position of

Jupiter. The right hand position is when Jupiter is moving towards the 180-degree position and the left-hand position, is when Jupiter is moving towards the same direction to be in line with Saturn. This position of opposite, at 180 degrees apart, and that of being together, as a conjunction, appear as significant. They indicated a connection to the changing of polarity of the main solar field and to the changing of the sunspot polarity. This means that the changing alignment of Jupiter and Saturn indicates at what point the solar field begins to change. We will consider these points again, further on. In an ideal situation, the two planets would always be ninety degrees apart at sun spot maximum but as we shall see this seldom occurs. It is important to accept that the repeated consideration of effects from planets at ninety-degree alignments is only a working hypothesis drawn from different observations. The ninety-degree alignments are not clearly apparent in the overall diagrams for the times of sunspot maximum. Nevertheless, Jupiter and Saturn appear to have a unique role and the focus is mainly on those two planets.

Both planets are of course continually moving and unlike a clock, the short arm is the faster one. We can consider this as a clock and inquire what makes the bell ring. That is to say, we can examine how the mechanism matches the main changes in the solar activity. From a visual viewpoint, we can consider the diagram as a right-hand pattern with Jupiter on the right and a left-hand pattern with Jupiter on the left. The terms left-hand and right-hand are a convenient mode of reference to the overall alignments in this study.

Alignments at Changes of Sun's Polarity

The two planets are in line, or exactly opposite each other some time after the maximum activity. This is usually three or four years later according to the positions at maximum activity. From investigations, my conclusion is that the change in the solar field relates to alignments of these two planets. That is to say when they are opposite, at 180 degrees or when they are in the same direction with no angular distance between them. The 180-degree alignment is an opposition and the second alignment is a conjunction. These terms remain throughout this study. It is the point of opposition or of conjunction that appear to relate to the changes of the main field and the sunspots. According to astronomical observations, there is an unusual situation at such times. There is actually a period where both poles can temporarily be the same polarity. Knowledge of all this developed in the mid-1950s, but backtracking carried out by means of advanced technical research methods confirms the phenomenon.

Although scientists did not generally consider the planets, some investigations helped to confirm my conclusions. As already indicated the average angular distance between the operative pair of planets is 90 degrees. This was further supported by a mathematical graph presented in *Solar and Terrestrial Influences on Weather and Climate*, (McCormack and Seiga, 1979). The graph showed the double coincidental positioning of both pairs of planets. That is, of Jupiter and Saturn as well as Uranus and Neptune. The authors emphasised the average 90-degree angular distance in both pairs but offered no explanations. As already noted there would presumably be an ideal configuration. If so, maximum solar activity would be after the time when Jupiter and Saturn were exactly 90 degrees apart. Minimum activity would be nearer the opposition or conjunction, and the magnetic changes would become complete as the planets moved past the line up positions of opposition or conjunction. The main phase would appear to be from the ninety-degree position to the alignment of the two essential planets, either as a conjunction or as an opposition. This difference stands out as the indication of why there is a double cycle of twenty-two years.

Ideal Planetary Positions

In this study, it seemed that there might be an ideal position for all the planets, which could indicate other key effects. The diagram, with all the planets in a sequence of 90 degrees apart, shows the ideal positions. This is in Figure 3. The distances of the planets from the sun are proportionate. The diagram shows that they all made very close 30/60/90-degree triangles so that there was an ordered "right-hand pattern". That is to say, with the sun at the focus, the planets in sequence make regular approximate 90/60/30 degree triangles. Using the same principle in a different manner, with the sun at the focus of a 60-degree angle, there was still a regular pattern. Either way there was a unique arrangement whereby the hypothetical planet in the asteroid belt also fitted in. From this, it was possible to estimate an approximate orbital period to see if it was a match in the cycles. The result was inconclusive, but this perhaps supports the scientific theory that the asteroids are the remnants of a planet that disintegrated or did not form properly.

The Question of Timing

Because of the variations in the actual angular distances between Jupiter and Saturn as shown in Figure 2a it seemed that we needed a finer timing mechanism. The astronomical clock perhaps needed a "second hand" to give the final time. The plant Mars appeared to be a likely choice, but Mars made no repetitive pattern at the main times. The table below shows the line up times and the angular positions of Jupiter and Saturn, ie. when they were opposite or in conjunction with each other. There are slight variations between the actual dates as the decimal expression as .1 of one year is 36 days. This also applies to Figure 2a, but the principle is clear.

The principle is that a new cycle actually starts when the two main planets are in line, as a conjunction or an opposition. However, there is a delay in the effect so the minimum follows some two or three years later. The maximum effect is when the two planets are ninety degrees apart, but other angular effects affect this and the delay in the activity. After that, the two planets move to their line up position and the process starts again. It seems that the polarity change would take place when the planets are in line.

TABLE 1–SUNSPOT KEY TIMES AND POSITIONS OF JUPITER/ SATURN AT TIME OF LINE UP

Cycle	Minimum	Maximum	Min to Min	Line up	Years between	Jupiter Degrees	SaturnD egrees	Alignment
13	1889.6	1894.1	12.1	1901.	-	285	285	In line
14	1901.7	1907.0	11.9	1911.	9.7	219	39	Opposite
15	1913.6	1917.6	10.0	1921.	10.1	176	176	In line
16	1923.6	1928.4	10.2	1931.	9.5	103	283	Opposite
17	1933.8	1937.4	10.4	1940.	9.9	42	42	In line
18	1944.2	1947.5	10.1	1951.	10.7	3	183	Opposite
19	1954.5	1957.9	10.6	1961.	9.7	53	53	In line
20	1904.9	1968.9	11.6	1971.	9.8	231	51	Opposite
21	1976.5	1979.9	10.3	1981.	10.1	187	187	In line
22	1986.8	1989.9	9.6	1990.	9.3	111	291	Opposite
23	1996.4	2000.2		2000.	10.0	52	52	In line

The table does not show the irregularity relating to 1883 but Figure 2a indicates that the pattern is irregular, assuming an alternating rhythm. Hypothetically, the maximum would always be just after the ninety-degree alignment of Jupiter and Saturn, with modifications caused by other planets, especially Uranus and Neptune. The indication is that technically the alignments of Jupiter and Saturn cause the changes but for practical purposes of observation, we observe the external appearance of the sunspots. As the change over of the solar polarity apparently takes some lengthy time, we cannot easily use this as a point in time.

All the planets are on elliptical orbits but whether this change of position has any effect on the actual sunspot activity is not clear. Overall the cycles vary from just over seven years to about seventeen years, but nine to twelve years seems more usual. In the column of figures for the line up times the average appears to be a steady ten years. However, it maybe that this is just a general mathematical repetition and has no significance. Nevertheless, whether we use the figures between the line up dates, or the figures between the actual sunspot minimum or maximum times, there is an irregularity in the rhythm, as there is a missing left-hand pattern in Figure 2a. The later maximum times after 1980 follow the general pattern and variations in sunspot relative numbers vary with angular contacts to Uranus and Neptune. Another irregularity in 1883, and 1894, was that the maximum occurred *before* the planets Jupiter and Saturn were at the ninety-degree angle. All the others in Figure 2a were *after* the 90-degree positions. This indicates that there is sometimes a different effect from the optimum 90-degree alignment. This may be an extra help in foreseeing the time of the sunspot maximum. Despite this apparent visual indication, it may not be highly significant because there are so many variables.

Irrespective of these points we can draw some conclusions from a comparative analysis of the alignments and the sunspot numbers at the times of sunspot maximum. The most interesting indication in this is that the sunspot maximum in the year 2000 is not "normal" in terms of a clear Jupiter-Saturn combination. All these points can be demonstrated by means of statistical comparison, but this still leaves the question as to how the process actually operates. In offering explanations as to how it works, we have to draw speculative conclusions that may not always be correct. Nevertheless, the indications are that the process is demonstrable in terms of physics and the specialists would be able to replicate the process in a laboratory and clarify what actually happens. The following table is an example that the repetitive alignments seem to suggest.

The Pairs of Planetary Polarities

The study demonstrates that the planets operate in pairs. This suggests that they are positive and negative. According to scientists, the earth is mainly negatively charged. From this, we can make a tentative table of planetary polarity. Starting from the inner planets, we have the following;

TABLE 2–SEQUENCE OF POLARITY OF PLANETS

Hypothetical Sequence		*Alternate Sequence*	
Mercury	Negative	Mercury	Negative
Venus	Positive	Venus	Positive
Earth	Negative	Earth	Negative
Mars	Positive	Mars	Positive
Missing	Negative	Asteroid belt	Neutral
Jupiter	Positive	Jupiter	Negative
Saturn	Negative	Saturn	Positive
Uranus	Positive	Uranus	Negative
Neptune	Negative	Neptune	Positive

If we take the earth as negative, there are two possibilities. One is with the missing planet in the asteroid belt counted as a sequence in the polarity. The other is with that planet omitted. In the first list, Jupiter is positive and Saturn is negative. In the second list, Jupiter is negative and Saturn is positive. Either way the main pair, of Jupiter and Saturn, has to be complementary in polarity. This conclusion is from the observation that the left and right-hand patterns change and the solar polarity and the sunspot polarities also change when the two planets change their relationship.

This indicates that there is a connection. Whether Jupiter, as an example, is actually positive or negative is not certain from these tables. It may be that astronomers do know such details. If so this will help in understanding the wider process discussed later. As Uranus and Neptune also appear to be an operative pair, some clues could come from a consideration of them. My own view is that Jupiter is positive and Saturn is negative and that the right type of comparison with the polarity of the sun spots would clarify this.

The Effects from Uranus and Neptune

Uranus and Neptune are very slow moving. Jupiter takes just under twelve years to orbit the sun, and Saturn takes nearly thirty years. The outer planets take much longer. Uranus completes one orbit in approximately eighty-four years and Neptune takes twice that time. As the effects of the two planetary pairs overlap, there is a change in the rhythms involving a shorter and a longer cycle. This idea of a longer cycle is not new and is a "beat effect". This beat effect stretches over a period of just under one hundred and eighty years. These two planets therefore make a half cycle effect when they are opposite or in line. At such times, the sunspot activity is generally less irrespective of the alignments of Jupiter and Saturn. If both pairs of planets are either directly in line, as an opposition or as a conjunction, there would in this principle be very little sunspot activity. However, this view is based on assumed ideal conditions but there can be unique exceptions and ideal conditions rarely exist.

TABLE 3–SUNSPOTS AND URANUS/NEPTUNE POSITIONS

Eleven-year cycles	Date of Sunspot Maximum	Uranus/Neptune Angle/degrees	Sunspot Relative Number
High Peak	1778.4	93	158.4
Low Peak	1816.4	11	48.7
High Peak	1870.6	87	140.5*
Low Peak	1907.0	178	64.2
High Peak	1957.9	86	201.2

*(This is not the very highest; 1837 =146.9)

As a point of interest the Uranus-Neptune angle for March 2000 is 13 degrees, yet the expectation was a high maximum. Furthermore, diagrams show Jupiter and Saturn almost in line. In addition, these two planets have a near ninety-degree angle to Uranus and Neptune at that time. This further indicates an "irregularity" as the above table suggests that the peak would be a low one. The final sunspot relative numbers show the truth on this and confirm some "irregularities", despite the fairly high peak.

There are of course high and low peaks in the normal eleven-year cycles. In the longer cycle, there are a number of such cycles. As indicated there is generally a higher sunspot relative number when the outer two planets are near the 90-degree alignment. Also there is "normally" a lower relative number when these two planets are nearly opposite or nearly together at a conjunction, as they were in this period. The alignments are never exactly at the ideal positions, however Jupiter, Saturn and other inner planets appear to determine varying degrees of activity.

The Principle of a Solar Generator

There are various possible processes, which might explain the many coincidences discussed so far. The specific function that interested me was in the change over of the solar magnetic field. It generally matches the change over in the relationship of Jupiter and Saturn. For example, Saturn might be in a particular direction and Jupiter is moving towards that position. Then, when Jupiter reached that alignment, to make a conjunction, the polarity is further affected. When Jupiter is "leading" the polarity of the sunspot pairs changes. Eventually, at the opposite position, Jupiter passes the 180-degree alignment so that as a pattern, Saturn is now "leading", and the polarity situation is different. By analogy, this broadly suggests an electromagnetic process similar to the way a generator, or dynamo, works. Yet, the enormous distances would preclude an exact generator model. Jupiter is 5.2 astronomical units distant from the sun. One astronomical unit is the distance of our earth from the sun. This is 150 million kilometres so Jupiter is over five times that distance away and Saturn is nearly double that distance. The planet Neptune is over thirty astronomical units distant. A generator with an armature sweeping through such a field is somewhat improbable unless we consider the solar field as the armature. In that model the looped force field lines would act as the armature and sweep round past the planets. That was my first assumption, but the problem was that the earth apparently operates in a similar manner, and the earth field is not like that, due to the distortion of the solar wind. Nevertheless, the generator principle had some merit.

In a generator process, the armature sweeps through the field of the magnets. The two magnets have their north and south polarities complementary to each other. Kinetic energy turns into electrical energy. In a mechanical generator, either the field magnets or the armature can be the moving parts. The magnets have an optimum position to give the best effect. If we apply these principles to the "solar generator" we have the possibility of kinetic energy, which is defined as "relating to the motion of material bodies and the forces and energy associated therewith," (Webster's Dictionary). Without doubt, we do have the motion of material bodies. Within a mechanical generator, the armature or the magnets can be moving. By analogy, the planets are the moving magnets. Also there is an optimum arrangement of the magnets. The optimum arrangement of the planets is at 90 degrees of angular distance. At this point, the model breaks down because we do not have an armature, and there is the question of what type of force or energy could fit into such a model. The most likely answer is in terms of energy waves. Radio waves are a type of electromagnetic wave, yet some types of waves apparently cannot penetrate the lower atmosphere. Even so, radio telescopes indicate that planets do emit some type of radiation, and as already mentioned, Jupiter and Saturn emit a strong radio noise. The term "electromagnetic waves" is therefore used with some reservation, as it may be an incorrect definition.

Instead of a solar armature, perhaps the planets radiate some type of electromagnetic energy, in all directions. When the sun is at the focus of two such forces, there is an increase in energy if the planets are in the optimum position. The implied process is that the core of the sun is stimulated. Convection currents develop, which are in pairs because of the two planetary effects, and the effect is opposite in the other hemisphere. Now, to jump forward to earthquakes, I have stated that the planets are radiating energy waves in all directions. This means that planets are not only acting on the sun, *they are acting on each other.* However, I have to confine my interest to our planet because I am linking this process to seismic activity. Nevertheless, changes in other planets may occur when any planet is at the focus of such forces. As a simple example, which is easily capable of being tested, another planet, say Venus, would have a change of radio noise when it was at the focus of two such planets. The "music of the spheres" would be changing all the time as different planets moved in and out of the focal point of *any two planets.* It would be interesting to check the difference when two planets were of the same polarity; presumably, there would be variations.

My conclusion is that the sun is not a self-contained nuclear reactor that will eventually run down. In addition, it could not be a self-contained generator, as the earth appears to operate in a similar manner. This becomes more apparent when we apply the same principle to seismic activity. Whichever way it operates it shows that there is definitely an inter-related process.

Missing Sunspots in the Maunder Minimum

To end this section on Sunspot Cycles I have to acknowledge the challenge to my hypotheses by the phenomenon of the Maunder Minimum. This is a period of little or no sunspot activity. E.W. Maunder, a British astronomer, studied old records and concluded that there were virtually no sunspots between 1645 and 1715. A later study of tree ring growth confirmed this. The rings show annual growth with more growth at times of increased sun spot activity. Edmund Halley (1656-1742) also observed an aurora for the first time in 1716 after forty years of watching. The aurora of the Polar Lights appears with the entry of solar charged particles entering the upper atmosphere of the earth. The lights are more prominent at times of sunspot activity, so Halley's observation is a further indication that there was indeed very little sunspot activity in the seventy years before 1716. Such a period would include six or seven sunspot cycles. However, Maunder did not say that there were *no* sunspots and records based on analysis show sunspot relative numbers. Yet the activity must have been low because reports of weather conditions refer to extreme cold which was associated with reduced sunspot activity; in other words the sun gave out less heat. Although definite evidence of this is difficult to obtain scientists associate a long cold spell, from 1550 to 1850 with this period and say that such phases occur about every 500 years. Halfway in the period is 1700. This period shows up in other comparisons, as discussed below and further on. However, the specific connection is elusive. The dates of Sunspot Maximum near that time were 1693.0, 1705.5, 1718.2 and 1727.5. The length of the cycles was 8.5, 14.0, 11.5 and 10.5 years. The first two are somewhat irregular in length as the average is near to 11 years. Specific analysis is difficult, as we have no data for that period. The numbering system of the cycles only started in 1755 and more accurate knowledge of the sunspot cycles did not develop until later. This was after the discovery by an amateur astronomer, H.S. Schwabe, in 1843 that the sunspots appeared in cycles. Records from that time are therefore better.

The Influence of Comets

Figure 1 shows the similarity of two graphs for sunspot activity and seismic activity. With fewer sunspots, there should hypothetically be lower seismic activity for the 70-year period from 1645 to 1715. Although there are no world records for that time, researchers say that the end of the 17th century and the beginning of the next century was a period of strong seismic activity. This perhaps indicates a possible irregularity and may offer some clues. Yet, it does not explain why there was lower solar activity in that period. The midpoint of that period is 1680. So we can ask what event could have had an effect both before and after that time?

My original interest focussed on Halley's Comet. It appeared in 1682 and has an orbit period of about 75 years. The later dates of its appearance are 1758, 1835, 1910 and 1986. The low period of solar activity is 70 years. This is almost the full period of the comet. Could it have an effect in that period and no other period? There is no indication of less sun spot activity at other appearances; it therefore does not fit into a pattern.

The path of the comet goes out past the orbit of Neptune. It comes in through Jupiter's orbit on its way to the sun, so hypothetically it could have swept past all the outer planets if they were in that direction. References to Halley's work mention a gravitational effect of the planets on the movement of the comet. What we seem to need is an indication of an electromagnetic effect on the planets. If there were only one very low sunspot cycle that matched the closest position of the comet, it would be easy to imagine a connection. However, there were six or seven cycles and the comet was making the full traverse of its path in that period. In that length of time, Jupiter would have passed through the nearest part of the path of the comet at least six times. Saturn would have passed through it at least twice. Uranus and Neptune might have been in that direction all the time as the path covers a considerable angular distance. Yet there are no records of comets affecting planets; in fact, the reverse occurred when Comet Shoemaker-Levy 9 surrendered to Jupiter's gravitational field in July 1994.

In addition to all this, the path of Comet Halley is not on the same plane as that of the planets. On its outer orbit, it is inclined below the ecliptic at 162 degrees. If Neptune and Uranus were in that direction they would have been far above it. It therefore is very unlikely that there would be any effect from Comet Halley. In any case, in 1910 the comet passed between the earth and the sun and part of its tail brushed the earth. There were no observable effects (except the sale of comet pills to protect the credulous) and overall it reveals that Comet Halley has no effect on any planets.

However, before dismissing this, it is of interest that there are other comets, and they do not all behave in the same manner. For example, Halley's Comet has an elliptical orbit which allows a calculable return time. Comets on different orbits are difficult to predict and in many cases, we do not know their period. There may be a comet that fits the requirements, although there is still doubt as to whether a comet could have any effect. Nevertheless the period of low sunspot activity was 1645 to 1715 and halfway was 1680. Texts on comets mention the Great Comet of 1680, which is not Halley's comet. It appeared at the end of 1680 and is sometimes called "the comet of 1680/1681" as its appearance lasted into the beginning of 1681. Whether a more irregular comet was able to exert any effect is a matter for expert debate. If the Great

Comet did have some effect, it is difficult for specialists to determine how, because there are is no detailed information. It is a "whodunit" without any real suspect, and we have to admit that so far the "culprit" has escaped detection.

Another difficulty posed by the Maunder Minimum period is that in the first half of the 18th century there was strong seismic activity. As I have tentatively suggested that flare activity might help to affect the earth's core we have jigsaw pieces that do not fit. This is discussed further on.

Global Warming

The reports of colder winters during the Maunder Minimum period indicate global cooling and therefore imply that an extra strong maximum would create hotter weather. In this context, it would be interesting to compare any indications of earlier global warming with sunspot activity and the positions of the planets in terms of the sunspot cycles. Perhaps global warming is a cyclical phenomenon within this framework. The details discussed earlier, in relation to the cold spell during the Maunder Minimum may be related. There may be a pattern in the 90/180-year cycles that relates to this, or it may be that unique slow moving alignments also affect the earth's core and make the earth hotter. Alternatively, perhaps even longer cycles relate to more than our solar system. There may be no records of such details but science often finds ways of checking. Within our own system the further details on earthquake and volcanic activity may offer clues on this and the present sunspot maximum may be of extra significance in these areas. The alignments for the year 2000 are unique, as were the alignments in December of 1699. Chapters 9 and 13 consider this further.

A Fusion Reactor

There is one further speculative query about the effect of planetary influences. Science books compare the core of the sun to a controlled fusion reactor where the compressed plasma of ionised particles continually creates energy until it runs down. However, if the stimulation of the nuclear process is via incoming electromagnetic energy waves from the planets, it could only run down when the planets stopped sending their energy. If scientists could replicate this process in the laboratory, it would demonstrate the viability of the planetary energy system. It may operate in a different form to that suggested here, but the principle could be similar.

CHAPTER 2

SOLAR FLARES

The Planetary Code for Solar Flare Diagrams

In defining a planetary code, numbers identify the planets. They are in the list below. A planet may be (1) for Mercury or (8) for Neptune and so on. The number code here is not the same as the one used for earthquakes. Solar diagrams are on heliocentric coordinates and seismic diagrams are with geocentric coordinates and include the sun and the moon. The number codes shown here are therefore only for solar flares. (0 degrees = the first point of Aries).

1	Mercury	5	Jupiter
2	Venus	6	Saturn
3	Earth	7	Uranus
4	Mars	8	Neptune

- The tangent arrows indicate the "indirect traction". This is the area where a planet is either rising or setting, and the area of low tide in ocean lunar tides.
- South is at the top of the diagram as heliograph projections create a reversed image.

- All planets orbit anti clockwise when looking down at the sun from above the solar North Pole.
- The viewpoint in every case is from the earth. The planets are in heliocentric coordinates.
- Zero degrees are in relation to the position of the earth.
- The details are from official *Solar Terrestrial Data* and *Astronomical Almanacs.*

The Trigger Effect for Solar Flares

Solar flares are short outbursts of concentrated energy in which jets of charged particles discharge outwards into space. They have a strong electromagnetic radiation that can cause magnetic storms, disrupting radio communication on the earth. The flare activity noticeably increases with sunspot activity. This is because most flares originate from sunspots. Some radio communications engineers have suggested that the planets are involved in these solar outbursts but the mechanism of operation eluded them.

This present study was not originally concerned with flares. It was only in the search for the final trigger effect for earthquakes that it seemed there could be a connection. The conclusion was that the same processes operate with the sun and the earth. There appear to be two distinct processes, and they both involve the planets. One is the internal effect or "generator effect" described earlier, and the other is a gravitational tidal effect that acts as the final trigger.

To make this clearer it will first be helpful to explain the basic operation of ocean tides on the earth. This is also in the section on seismic disturbances. In ocean tides, there is a high tide mound of water drawn from one half of the planetary surface. The cause is gravitational force from the moon and from the sun. On the far side of the earth, there is a complementary mound of water although the moon is on the other side. This is a "negative bulge" of the ocean tide and is the reason why there are two high tides in every twenty four-hour period. With ocean tides, the main effect is from two bodies. These are the moon and the sun. In solar tides, we have to consider tidal effects from the planets. Ocean tides have two main points, one is at the position of high tide, and the other is at the position of low tide. With both solar flares and earthquakes, the most critical point is at the position of low tide. This is when another celestial body is either rising or setting. For the earth, it is usually the moon and the sun, but can be planets. For the sun, it is of course only planets.

This setting or rising effect is a type of "Indirect Traction". When the body is overhead, it is "Direct Traction". These are terms used throughout this study.

Planets and Flares in a Selected Period

The detailed solar flare study was for a period of low solar activity. This choice was deliberate, to reduce the risk of coincidence. The period chosen was for 1963 and 1964. The details were from official solar flare observation records. The selection was one hundred flares at intervals of not less than four days apart. This would then allow any sunspot to rotate out of the previous position. The reasons for this are complex. The sun rotates at different speeds at different latitudes. At the equator, it takes about twenty-five days and at higher latitudes, it takes over forty days for a full rotation. On an average, any point on the sun would traverse the visual disk in about fourteen days. Apart from this movement the planets are also moving. Mercury makes a complete orbit in 88 days, Venus takes 225 days, Earth takes one year, and Mars takes just over 1.8 years. (The other planets are much slower. There is consequently a changing rotational transit of the sunspots and the changing orbital velocities of the planets. It is from this continual play of forces that helps to create a solar flare. It is also of some interest that sunspots always first appear in higher latitudes of around 40 degrees north and south but can be higher. They then rotate round and down towards the equator where they dissipate.

This phenomenon is not crucial to our inquiries but may be of interest in considering why one spot should be disturbed and not another, for there are times when many spots are visible.

The study consisted of one hundred diagrams set out as summarised in Figures 5a, 5b and 5c. These have a plan view and a side view as seen from the earth. As already stated, north is at the bottom because solar-observatories use a reflecting system that gives a reversed image. In a mirror image East-West are reversed but for a Plan diagram East-West must be the same, so North-South are reversed.

Figure 4 demonstrates the principle involved. P1 and P2 are two separate planets and the extension to the tangents for each planet shows the limits of the gravitational field of each planet. The point where the tangent touches the circle is the low tide point. It is the position of Indirect Traction, and is either point where a planet would be rising or setting. In general, the flares occurred at this point, but it was more often when two fields were overlapping. The flares were therefore from a sunspot affected by two traction fields. This principle also shows up in seismic disturbances. I refer to it as "contrary or dual traction" and see it as an area of conflict where the surface is trying to move in two directions.

At this point, the gravitational force of the planets is not pulling directly against the solar mass but is in effect bypassing it, because the traction is at a tangent. This position, where the sunspot rotates into or out of the gravitational field of the planets is where there is a disturbance. As mentioned earlier it may also be the case that there is a direct radiation effect from the planets. In other words, the gravitational field could be the radiation field for any particular planet. It is likely that there is some direct electromagnetic radiation because radio telescopes, on the surface of the earth, are able to record a radio noise from a planet. Whether this is of a sufficient intensity to have an effect is not certain. The open question is whether there is gravitational disturbance or electromagnetic disturbance. Either way the overlapping fields show up as critical. For the purpose of discussion, they are gravitational fields whereby the disturbance is by traction.

Figure 5a is for an actual flare and is a typical diagram. Out of the one hundred flares checked, seventy-five were on or near the edge of the overlapping area as shown in Figure 4. Six were doubtful, as the overlapping area was somewhat large. Four were in the correct position but the two planets were pulling in the same direction. Fifteen were negative and apparently not related to the principle described. The general pattern was that planets were either just rising or just setting on the point of disturbance. The same type of pattern showed up with seismic disturbances but a stronger combination is then necessary. Two typical examples are in Figures 5b and 5c. Figure 5c shows the position of two flares at slightly different latitudes. This indicates that the same forces affected two different areas of the sunspot. The total area of a sunspot is enormous so this is quite likely.

An interesting phenomenon that emerged during this study was that there would be a disturbance on the Western limb, right on the edge, then it would be quiet where there were no traction effects, and there were no flares recorded. Yet, when that same area reached the other limb and a suitable traction pattern existed for that limb, there would be a flare. In addition, by observing the shifting position of the planetary traction zone it was noticeable that the flare was then earlier or later in relation to the changes. For example a flare on the Western limb as in Figure 5c suggests that there will be no further disturbance until the sunspot again rotates into the traction zone. Of course this would have changed with the faster moving planets but would almost be the same with the slower ones, such as Uranus and Neptune. In sequence, a series of diagrams demonstrate this. This principle suggests that observers could anticipate flares and solar observers could plot the likely time of the flare. Observers could perhaps foresee flares likely to strike the earth.

Flares Related to Planetary Positions

A detailed breakdown of the results is below. This relates the planets to the flares. In general, the flares occurred when only two planets created an overlapping field. With two planetary fields overlapping, it is noticeable that the outer planets were more often involved. This is because the slower moving planets would be making an overlapping area for a longer period. Sunspots would therefore pass through that area more than once. The ratings are in Table 5.

Further details of the relationship of planetary alignments to solar flares are in Chapter 11, dealing with the sequence of effects. Overall, we have to consider the coincidences, the mechanism and the sequence. At this stage, we are mainly considering the coincidences in positions of the flares in relation to the planets.

TABLE 4–DETAILS OF FLARES AND TRACTION

Position	Number of flares.
Edge of overlapping traction area.	38
Inside overlapping traction area	25
5 to 10 degrees outside overlapping area	12
Total	75
Number outside traction areas	25
Total overall	100
Differences in amount of overlap of fields.	
Greatest overlap	74 degrees
Smallest overlap	7 degrees
Average overall	36 degrees
Variations in sunspot latitudes	
Highest latitude	60 degrees
Lowest latitude	7 degrees
Average overall	19 degrees
Number of planets creating overlapping field	Number of flares
2	36
3	27
3 +	12
Total	75

TABLE 5–PLANETS RELATING TO FLARES

Planet	Flares in field
Neptune	15
Uranus	13
Saturn	9
Jupiter	9
Mars	5
Earth	8
Venus	6
Mercury	4

Planetary Alignments and Sunspot Relative Numbers

I particularly studied the likely effect of other planets as a means of ascertaining whether any two planets at right angles to each other might have an effect as with seismic activity. Hypothetically, the appearance of sunspots would follow a similar pattern as that for seismic activity. The suggested principle, as already described, is that planets at right angles to each other affect the core of the sun or planet at the focus of those two planets, where the radiation forces intersect. This apparently stimulates the core and leads to easier surface movement. Comparison with helio-centric planetary alignments and sunspots gave indications that there is not an instant response. It therefore seemed unlikely that there would be an immediate effect on the earth. This was easier to observe by following current reports. From observations of the times of seismic disturbance, there is some delay. However, the effect may begin to take place before the exact alignment so it is difficult to pin point an exact time. In general, there was an increase in sunspot relative numbers when any two planets were at or near 90 degrees apart. The following analysis is from a study of 1963 records. The last column on the right shows the sunspot relative numbers for two dates to demonstrate the increase. The sunspot relative numbers indicate the degree of sunspot appearance.

From these figures and from the analysis described earlier we can see that the sunspot maximum is most likely to be when as many planets as possible (including the earth) are making ninety-degree alignments with the sun at the focus. This is when the core of the sun is most active, but Jupiter and Saturn are the deciding factors, because they determine the change of the magnetic poles. Most flares will be associated with more sunspots and more indirect traction from the planets disturbing the sunspots. The tables and diagrams help to demonstrate this effect.

TABLE 6–PLANETS AND SUNSPOT RELATIVE NUMBERS

1963 date of main activity	Planets at or near 90 degrees	Dates and relative numbers				Increase
January 4th	Venus - Neptune	1st	23	4th	35	12
January 15th/17th	Saturn – Neptune	13th	9	15th	44	35
February 4th	Earth - Saturn & Mars - Neptune	1st	36	4th	53	17
March 8th	Venus – Mars	4th	14	7th	35	21
April 7th	Earth – Venus	4th	17	8th	63	46
April 15th	Earth – Venus	9th	48	12th	63	15
May 7th	Earth – Saturn	5th	23	10th	64	41
June 7th	Mercury – Venus	6th	8	12th	87	59
July 3rd	Venus – Uranus & Mercury-Saturn	June 30th	27	3rd	37	10
July 31st	Mars – Saturn	29th	24	31st	65	41
August 18th	Mercury – Uranus & Venus - Mercury	15th	11	18th	43	32

August 22nd	Mars – Saturn	20th	36	22nd	68	32
September 4th	Mars – Saturn	1st	20	4th	43	23
September 15th	Mercury – Mars	12th	28	15th	84	52
October 7th	Mercury – Earth	7th	9	9th	37	29
October 11th/ 14th/ 15th	Mars – Uranus & Mercury – Venus & Venus-Saturn	10th	32	17th	50	18
November 15th	Mercury – Uranus	14th	7	23rd	36	29
December 2nd	Uranus – Earth & Venus - Neptune & Mercury	Nov 29th	21	2nd	31	10

From the above details, it appears that any two planets could affect the core of the sun or of the earth. This is by stimulating core activity, but there would be no flare, or earthquake, unless the tidal effect also acted as a trigger. Alternatively, observation shows that there can be a surface disturbance without the appropriate tidal effect. Even so, there is not necessarily a surface disturbance when both processes appear to apply. My own conclusion is that there may be other factors, apart from the planetary effects, and these could explain the apparent exceptions to the main principle.

Finally, there is one point that showed up in seismic diagrams that may be useful in determining when a flare could occur. This is the distance between the opposing traction limits. There may be an optimum area of overlapping forces. Say, for example, on the diagram the disturbance showed up when the traction limits were n distance apart. This means that the traction limits may be n distance apart in a high latitude but perhaps a greater distance apart at the equator. The flare might then be in the high latitude in the narrower area. Alternately the distance between the traction limits may be even less at high latitude and then proportionately narrower at the equator. In this case the flare would be on the equator, that is, if there were such a "rule". Although there would be differences from the different traction of the planets, it should be mathematically calculable as to where the opposing forces have the strongest effect.

In applying these principles to seismic activity there were various similarities that appeared useful in attempting to anticipate a disturbance. The most noticeable was that the peaks in the graph of earthquakes definitely coincided with major planets at right angle alignments. Strong ninety-degree angles clearly related to extra seismic activity, and the absence of such alignments matched a drop in activity. The indication is that such alignments do affect the core of the earth, (and the sun), and that this is the initial requirement in the build up for the gravitational forces to trigger the final effect. Further details on this are in the next section on earthquakes.

CHAPTER 3

PLANETS AND EARTHQUAKES

Earthquake Planetary Code

As already explained, the number codes for the earthquake diagrams (Figures 8 to 10) is not the same as for the solar flares. This is because we also have the sun and the moon as external bodies. The number code is below.

For the earth, the coordinates are usually on a 24-hour scale referred to as "right ascension". This relates to the 24-hour rotation of the earth on its axis but the datum point of zero degrees Aries is still the starting point. In geometric terms six hours equates with 90 degrees and so on. One degree is equal to four minutes so converting right ascension coordinates to a measurement in degrees is quite simple. The solar heliocentric coordinates are always in degrees similar to navigational almanacs whereas planetary positions from the earth are in right ascension. Apart from observatories most large city libraries have the official Astronomical Almanacs, including back issues to the beginning of the century or earlier. Students should have no difficulty in obtaining the data, especially as there are now CD-ROMs available which provide geocentric and heliocentric coordinates.

Details of earthquakes are obtainable direct from the Department of Geological Survey or on the Internet. Geophysical Observatories also have copies but do not supply such reports. I originally obtained monthly seismic reports direct and also obtained the monthly flare listings or for individual years as required, from the National Bureau of Standards USA. More recently, I have used the Internet to access the Northern California Earthquake Data Centre (NCEDC) figures.

EARTHQUAKE DIAGRAMS NUMBER CODE.
0 degrees = the first point of Aries.

1 Sun	6 Jupiter
2 Mercury	7 Saturn
3 Venus	8 Moon
4 Earth	9 Uranus
5 Mars	10 Neptune

The number code is arbitrary, and does not rank the traction effects of the planets by strength. It is included as a means of showing the relevant 90-degree alignments. Originally I considered a traction effect in which Uranus and Neptune were insignificant, but they cannot be completely ignored. The moon has the strongest tidal effect, whereas Mercury only has a minimal effect. Its significance as a mass, discussed later, shows a unique result.

As with the solar diagrams the tangent arrows indicate the indirect traction and show where the planets are rising or setting on the seismic area.

The North Pole is central, for convenience. There is therefore no allowance for the inclination of the earth in the general diagrams unless specifically shown. However, this creates an error relative to the latitude. Nevertheless, the diagrams show the principles involved. Tables of rising and setting times are the simplest method of making a correction.

The terms "conjunction" or "conjunct" and "opposite" and "opposition" are the same as for the solar diagrams. Some of the following details were included in my earlier paper (1979).

Planetary Effects with Earthquakes

In applying the principle of planetary effects to the earth there are some differences. The central process is that first two planets stimulate the core of the sun by the electromagnetic process, and then two or more planets disturb the sunspot by gravitational tidal forces, causing a flare. The main difficulty for earthquakes is that no specific planets show up as significant. Therefore the matching of the seismic graph with that of solar disturbances does not mean that the two key planets of Jupiter and Saturn are equally responsible for seismic disturbance as well as solar activity. The increase of seismic activity with the increase of sunspot activity also appears linked to the increase of solar energy reaching the earth. That is to say, more solar particles perhaps help to activate the core of the earth and make the mantle more plastic. This alone would then help the tectonic plates to slide more easily. It therefore might be the reason why there are exceptions to the main principles. As well as this, the amount of traction needed to affect a seismic fault varies. This probably relates to the extent of heat stimulation affecting the mantle. That is to say, when the mantle is more fluid it requires less traction to move the fault. Also the surface of the earth is not as uniform as that of the sun; some faults may be more difficult to disturb. Faults on the east coast could receive horizontal compression from ocean tides that are sweeping westward, because the earth is rotating eastward.

Saying that earthquakes on earth are equal to solar flares on the sun is only a broad analogy. The main difference is that the surface of the earth is different and the fault lines are fixed. Added to that we have regular ocean tides caused by the nearby moon. However, there are the same two main effects. One is the effect on the core. The other is the effect on the surface. On the earth, there is also the effect of ocean tides. In addition, there is an added phenomenon on earth; it has biological life. There is therefore a possibility that the associated electromagnetic effects may affect such life. We will consider this separately.

The basic position appears to be as follows;

- Extra solar energy from solar radiation increases seismic disturbance by helping to make the mantle more fluid.
- Any two planets at a suitable alignment can add to this by stimulating the core of the earth.
- External traction from the sun and the moon and the planets can then affect the fault because the mantle is more fluid.
- There is also an effect from ocean tides.

Of course the plates are not stationary and only moved by external traction. They are moving all the time. This is mainly from the rotation of the earth that tends to make the plates "drag" and slide westwards because the earth rotates eastwards. In addition, the ocean tides repeatedly thrust against the east coast of all landmasses. The tides move westwards by trying to stay under the moon that is causing the tide or mound of water. In fact, the tides do not move at all. It is only the earth, rotating eastwards, that creates the appearance of the tide waters moving west.

Geophysical Investigations with Earthquakes

Modern specialist literature on the subject of geophysical disturbances shows a very broad field of investigation. It ranges from consideration of the ionosphere right down to the centre of the earth. Each area is enormous in its ramifications and subjects such as oceanography are separate fields of study. We can only touch on the basic principles in this attempt to show how the planets come into this intricate picture.

Concerning seismic activity, geophysicists tend to look to internal developments as the main cause. They link volcanic activity with seismic activity and see both as an extension of activity in the outer core and in the mantle. Texts on the subject explain that heat is an important factor in the process. Specialists say that pressure or radioactivity is the cause of this heat that then affects the upper mantle. The magma is then softer and the tectonic plates can drift more easily until pressure builds up and the crust eventually slips. The suggestion of radioactive elements in the earth is significant. There is little doubt about this as the release of radon gas near a fault indicates a radioactive change. The mantle is therefore not like a fluid that is heated from below but is radioactive itself. All this is well known and there are many books on the subject. Nevertheless, the actual cause of the heat is something of a mystery. Also there are variations in the heat. What causes the heat and what causes the variations is part of our inquiry.

As already explained the influx of solar charged particles at the poles might cause some heat. This increases slowly as the sunspot cycle progresses. We then have the specific questions as to what causes any extra heat and short term fluctuations. The indication is that it may be by radiation from the planets. We then have the question as to what is the final trigger effect that causes the earthquake. Before we consider that a few more background details for the study will be useful in understanding the overall process.

Planetary Positions for Earthquakes

As a basis for the study, I selected the fifty years, from 1900 to 1950 in *Seismicity of the Earth and Associated Phenomena* (1954) by Gutenberg and Richter. This work included all major earthquakes in that period, the majority of which are above five on the Richter scale.

The exact location and the Greenwich Mean Time (GMT) or Universal Time (UT) are given. Together with the *Greenwich Observatory Astronomical Almanacs* it was an ideal source of information. From the lists, I made a selection of fifty. This averaged about one major earthquake from each year. I later selected a second set of fifty to see if the same results occurred. In the meantime, I carried out another study. The graph in Figure 1 is from the earthquakes listed in Gutenberg and Richter's lists. However, I wanted to make a check to see if a graph of all earthquakes, irrespective of magnitude, would give the same type of graph. I inquired about this to the Geological Society, in London, as I was living in England at the time (1963). They offered to send me full records of all earthquakes in that period and said I could have them for one month if I paid for the freight. I agreed to this and a railway delivery truck eventually rolled up with a large packing case containing the reports.

Fortunately, I had a team of "workers", so I took all the material to the school. As the initial graph was the result of a class project, the students were not too surprised when I explained the situation. It was really only a counting exercise. We decided to count on a monthly basis. For fifty years, it gave us six hundred points on the graph. We reduced that to a workable number. The students then had a one-month project in counting and making a graph for a fifty-year period. We did it but we need not have bothered. The outcome was that the overall graph was almost identical with the graph of major earthquakes. It would have been strange if it had not matched, but while I had the records, I copied out details of many lesser earthquakes. Later I checked them against my hypothesis that external forces helped to trigger earthquakes. Again the results were similar but there were variations and frustrating missing parts of the jigsaw.

The Role of Planets in Earthquakes

In the beginning, the search was for any repetitive planetary alignments that coincided with seismic disturbances. The second stage was a search for evidence of any external bodies either rising or setting on the disturbed area. Side by side with this was the study of planetary alignments at times of solar disturbances. As already explained the situation with sunspot cycles was easier because of the repetitive arrangement of the main planets in definite pairs. There appeared to be no definite firm pattern of alignments with earthquakes. Nevertheless, in general there were at least two planets at or near a 90-degree angle at the time of the earthquake. The pattern was similar to the list given at the end of the section on flares.

In this check, I ignored Mercury because it is too near the sun. In addition, Pluto seemed unimportant. In the consideration of gravitational traction, Pluto is remote and has a small mass. It seemed there would be no calculable effect, although Pluto may emit some radio disturbance and relate to the core disturbance. At the time, I could not find any literature dealing with radio telescope examinations of Pluto.

The initial conclusion was that the main core stimulus could be via the sun, from solar particles that entered the earth at the poles. This is because the Polar Light Aurora fluctuates with the incidence of flares. However, it seemed that a secondary independent and parallel stimulus is via the planets. The indication was that if both the planets involved were slow moving there would be a longer effect. With faster moving planets, the effect would be more noticeable. This does not mean that both the planets would be faster moving. Only one of them needs to cause the pinpoint in time when the optimum angle is completed.

Out of the total one hundred earthquakes, the following details emerged.

TABLE 7–EARTHQUAKES AND PLANETS NEAR 90 DEGREES +/-
10 DEGREES

Number of times individual planets were involved	
Neptune	4
Uranus	4
Saturn	3
Jupiter	3
Mars	2
Venus	2
With more than one right angle alignment	4
Total number of shocks within those limits	8

Of the planets outside the limits of plus or minus ten degrees a number were only a few degrees less than eighty degrees or more than one hundred degrees. They were not included because the search was for a specific limit. However, it is relevant to point out that with solar disturbances Jupiter and Saturn were never exactly at the optimum angle. The difference there could be because other planets added to the effect.

In considering these facts, it may be relevant to repeat that this process is happening all the time to all the planets. It therefore may be that in their role as both receivers and transmitters of energy that they vary in effectiveness. As already stated this might show up with radio telescope investigations.

The External Trigger for Earthquakes

The final trigger effect for earthquakes shows a relation to external gravitational forces. Of these, the sun and the moon are the most prominent, with Venus and Jupiter next. A grouping of planets appears to add to this effect especially if both the sun and the moon are in the same direction. It seems that this external traction will give the final jerk and disturb the fault. This jerk may be as the fault enters the traction zone or gravitational field of the external bodies, or when it leaves it. Geophysicists refer to an "elastic rebound" effect with earthquakes but do not fully explain it. The indication is that the crust springs back to its original position when it escapes from the external traction. These points are in detail further on.

In general, the external traction is not from a direct source where the external body is overhead but is indirect. That is to say that it applies at a tangent to the point of disturbance. This is the point where the external body is just rising or setting. As the moon is noticeably effective in this process, it is also at the point of low tide. Scientific comparisons with tidal effects focused on the time of high tide but a comparison with low tide times is of equal interest. Although some of these points are complex, it is of particular interest to understand the difference that showed up in such investigations.

The very simplified "rule" is as follows. A large landmass experiences more earthquakes when the external bodies are either rising or setting, especially with the moon. Islands are more affected when the external bodies, including the moon, are overhead. In simple lunar tidal terms large landmasses are more affected at low tide positions, and islands are generally more affected at high tide positions.

The reason is that at high tide, for an island, the area all round the island has an extra weight of water caused by the high tide. The local shelf then has more weight on it. With a large land area, there are also earth tides so the traction principle is generally applicable. As already mentioned faults near the coast can be stressed by horizontal compression caused by ocean tides, or by springing when there is less water at low tide. The moon causes these tidal effects so we are still dealing with the effect of traction from external bodies. As an aid to clarification, the principles in ocean tides are now further discussed.

Effect of Ocean Tides on Earthquakes

We have already considered some details of ocean tides. They are discussed here in more detail because not only is it an important aspect of the process it is also very complex. The high tide is really a mound of water caused by the attraction of the moon, which is approximately overhead. At low tide the moon is just rising or just setting on the low tide position. However, there are some variations and a study of daily moonrise and moonset times and tide times in the daily newspaper of any coastal town will show differences. The sun is the cause of the main differences. The sun has an enormous mass but its gravitational effect is less than that of the moon because it is further away. The effect of these two bodies would be simple if the earth only had an ocean, but it does not. Consequently, the landmass often blocks the tidal movements. This causes horizontal compression of the crust when the water increases on the continental shelf and moves into bays with no outlet. This helps to stress any faults in such areas. This horizontal compression may be right across the land, such as in North and South America and the effect is then nearer the West Coast than the east coast. Mountains on either coast are the result of such pressures and relate to nearby faults.

The highest tides are at New Moon, when the sun and the moon are overhead. A very high tide is also at Full Moon when the sun and the moon are on opposite sides of the earth. At such times, it is a Spring Tide but there is also a Neap Tide caused by a regular positioning of the sun and the moon. This is at the first and third quarter of the moon phase and is when the sun and the moon are 90 degrees apart. However, this 90-degree position is not the 90-degree process discussed in relation to planets. The sun and the moon do not affect the core of the earth in a similar process but only affect the ocean tides. Because the sun has a considerable effect on the water the Neap Tide mound is not directly beneath the moon as at a New Moon or a Full Moon time but is somewhere between them in angular distance. It is not exactly at the halfway mark of 45 degrees because the moon is exerting a greater pull on the waters. Broadly speaking it is always nearer the longitudinal position of the moon than that of the sun. From this it can be seen that the Neap Tide "bulge" of water is only in a temporary position and is continually shifting to adjust its relative position to the moon and the sun on any given day.

There is also the problem of the complementary tide on the other side of the earth. In terms of tidal horizontal compression, such pressures are occurring twice every day. The fault is therefore under stress by horizontal compression *from the complementary tidal bulge.* The simplest way to resolve this is to consult tables for high and low tides because they include the tide times for both aspects of the tide mounds that are sweeping round the earth.

The effect of this continual pressure stresses the continental shelf as the weight of water moves off and on to it. The series of blows from this moving water is like the a line of heavy ball bearings where a strike at one end causes another at the far end to move while the ones in between remain still. This is the same principle in hydraulics whereby the force of a blow at one end of a pipeline travels to the other end. The impulse carries through so that a high tide on the east coast may help to cause an earthquake on the West Coast. The point here is that there are two effects. One is the changing weight of water on the shelf. The other is the horizontal blow on the coast.

Apparently, New York experiences many minor tremors but there have been no major quakes there in recorded history. There is of course no major fault there but it indicates that there is some pressure. The effects are probably nearer to the time of high tide than the low tide.

Planetary Forces with Earthquakes

Apart from all these complexities, there are planetary tidal forces. These are not a factor in ocean tide predictions and apparently, tide tables are from an analysis of past observations. Nevertheless, there appears to be a planetary effect with a seismic fault under pressure. An analysis of the diagrams showed that not only were planets often rising or setting but sometimes there was a combined effect from planets pulling in a near opposite direction. With two contrary planetary effects, the angle that appeared most prominent was approximately 140/145 degrees. It matched the type of overlapping fields that showed up with solar flares. So for earthquakes, there are two traction patterns; one pattern shows external bodies in the same direction and the other with overlapping gravitational fields. With solar flares, the overlapping fields were more prominent but with earthquakes, both types appeared very frequently. However once a fault has been disturbed less external traction could affect it. This shows up as aftershocks.

The principles involved are in the following diagrams. Figure 6 shows the principle involved with only one external body. The letter "E" shows the external body in the line of direct traction. The point designated by the letter "I" is the area where the crust of the earth is either just entering or just leaving the gravitational field of the external body. As the earth rotates, every fault would experience some stress when it passed into or out of this field. It suggests that there would be a slight jerk forward as it entered the field or a slight rebound jerk as it escaped from the traction. At the point "I", the dark section in the diagram indicates a slight gap as the plates move sideways. The darkened segment in the line of Direct Traction indicates how the crust rises towards the external body. There are reports of the crust lifting as much as thirty centimetres. This passes unnoticed because it is gradual. A simple model using a section of a cylinder and with segments of metal loosely touching and secured with an elastic band shows that all the segments move when a strong magnet is near. The nearest segments tend to move outwards but the others *slide sideways*. However, a tectonic plate stays in position because of adjacent plates and the gravitational force of the mass of the earth but it can slide sideways. Nevertheless, as the earth is rotating this creates a conflict of tensions between the sideways pull and the rotational force of the earth. This would appear to be different at each point "I"' as one is entering and the other is leaving the traction zone.

In Figure 7, the four diagrams show the different details of the external traction. The first diagram (a) shows the ideal effect of the moon at high tide with the moon overhead and no other external bodies adding to it. The low tide positions are where the water has moved away because of the external traction. In three-dimensional terms, the high tide would be a mound of water. Other forces and conditions modify the effect in real situations. The second diagram (b) shows the tangents that create indirect traction to the low tide area. The third diagram (c) shows that the traction is over one half of the earth. In three-dimensional terms, this means it is over one half of the sphere of the earth. The next diagram (c) is a side view of the earth and shows the traction limits. The last diagram (d) shows dual traction from two external bodies. It shows an overlapping area where the two fields are on the same area and opposite is an area that is not affected. This may be a relatively safe area. There are times when all the planets, the sun, and the moon are involved with such a pattern of cross traction. Dark areas show the tidal effect. The triangular shaded area at the top of the diagram is where the earthquake often occurs.

Earthquake Diagrams with Planetary Positions

The first example is for the well-known San Francisco earthquake of 1906 (Figure 8). It is a good example of traction in the same direction. The tangent arrows showing the traction only plotted onto the earthquake area as a means of keeping the diagram clearer. In a full diagram, they would also be on to the other side of the diagram. The tangent point is always at right angles to the line of direct traction. (It is actually a great circle right round the globe). If we consider all the external bodies, we can see that in this case the general low tide from the traction was almost in the centre of the indirect traction from the sun and the moon. The impact was just before dawn, before the sun rose, as indicated by the tangent arrow (1). The earth is rotating anti-clockwise so the moon (8), had risen about three hours earlier and Saturn (7), was also well risen.

If these positions are plotted onto a globe of the world it will be seen that overall, the low tide area was right over into the Gulf of Mexico. The main high tide mound of water would be somewhere in the East Atlantic Ocean. Then the high tide mound would move into the Gulf of Mexico and unable to escape. There would be strong horizontal pressures. At a low tide position there would be less water on the east coast to act as a buffer for eastward moving traction on the land. The conflict of tensions involves ocean tidal forces and external traction. A study of tide levels in all that area might show variations from the norm. There would be a very high tide and a very low tide. The implication is that at low tide the continental shelf is more likely to spring as there is less weight of water to hold it steady. In addition, the crust is moving sideways so is not so restricted by the gravitational pull of the earth. In terms of mechanics, it might be possible to ascertain when such a play of forces might reoccur. Apart from practical observation of the planetary movements, there could be a possible analytic forecasting as with tides or weather.

The two planets at right angles to each other were Jupiter and Saturn. This is significant because they do have a very strong radio noise. Sunspot maximum was at the very beginning of 1907. However there is one very interesting and serious fact with such diagrams. They often showed Mars, Jupiter and Saturn aligned in the same general direction as the sun. This means that at that time they are on the far side of the sun. Their gravitational force would therefore be less than if the earth were at its nearest point to these planets. We will consider this problem of relative mass further as the relative traction depends on the angle of the line of force.

The second example is for a major earthquake in Iran (Figure 9). It is an example of cross traction. It occurred just before the Full Moon and it was the time of an eclipse of the moon. This is when the earth would be exactly between the sun and the moon. The two gravitational fields would then coincide almost exactly. The time was more than an hour after sunset and the moon had already risen. The cross traction is from Mars and Venus. In general, Venus and Jupiter are often prominent in the diagrams because of their relative effect in terms of mass and distance. Mars is at times relatively close to the earth but in this case, it was almost on the other side of the sun. Yet, it might figuratively help to add the last straw. As just mentioned, we will discuss this later in terms of Mercury's small mass.

It may be that in the last analysis a simplified consideration of external traction from the planets is not adequate. Mars and Venus are often prominent in these diagrams. In the selection, they were often within 60 degrees of each other but there may be other effects. However, gravitational traction appears to be the general case, so that is my focus.

Because of the position of the sun and the moon, the high tide water mounds would be in the mid-Atlantic Ocean and moving westwards in the Pacific Ocean. A study of geographical features on a globe shows the build up of mountains right across from China to the Zagros Mountains west of Iran. The significance of this is that horizontal compression helps to create mountains and the pressure of ocean tide helps to cause horizontal compression. The point is that the trigger may apply to the seismic area but the horizontal pressure may have come from some distance away. We are also here considering earth tides as the crust may be pressured sideways by the ocean tides and pulled sideways by the traction forces applying to it. This may not however be in the same direction as the ocean tide pressure. It may be helpful here to reconsider the overall problem by recalling that we are concerned with three forces in this process. The first is the general prerequisite of two planets at an alignment that affects the core of the earth. In this case, Jupiter and

Venus were at a very close right angle position. As already mentioned these two planets have a strong radio noise. The second force is the horizontal compression, caused by ocean tides. This may have been building up for some time, and finally there is the trigger effect from planets or the sun and moon setting or rising on the stressed area. However, the diagrams shown here are to some extent ideal examples. In some cases, the exact analysis is very difficult.

Uranus and Neptune were not originally included in the diagrams as I thought their great distance would offer no significant gravitational effect. In addition, I was not sure of any effect on the core of the earth. The later flare study indicated that they could have an effect on the core as well as on the surface of the sun. They made no suitable angular connection for the San Francisco earthquake but Uranus was one hundred degrees to Jupiter and Neptune was 90 degrees to Mars at the time of the Tabas disturbance. This therefore made three near ninety-degree alignments affecting the core.

The third example is a more dramatic emphasis on ocean tide forces. It is for Mexico City in 1978 (Figure 10) about nine weeks after the Tabas earthquake. At the time of the shock, the traction forces were pulling to the west whereas with San Francisco they were pulling to the east. The sun and moon were close in alignment as it was near the time of a New Moon. Planets were also adding to their effect. It would have been a high Spring Tide at around midday and the earthquake was two or three hours later. My evaluation is that as the earth rotated eastwards the higher mound of water in the Atlantic Ocean would have tried to sweep westwards to stay under the sun and the moon. However, the land would obstruct the tide. The water would have flowed into the Gulf of Mexico and piled up on the shelf, unable to escape. The Panama Canal further south is unique in this problem. Water at one end is much higher than at the other, hence the number of locks to step down to the other level. The tide water trapped in the Gulf would also be under traction. We know that large dams have tidal movements and diagrams for shocks under dams often show traction onto the area. As the moon rises and sets the water at one end tries to move to the other end. With additional traction from planets, this would be more than usual. As with ocean tides checks of east-west water levels in dams would probably show how differences matched variations in external traction.

On the western side of Mexico the high tide in the Pacific would have recently been on the continental shelf and then moved off it as the earth rotated. The diagram suggests that the main group of external bodies were still pulling on the water that was to the east. It is a particularly good example of traction adding to the ocean tide forces. However, the initial effect on the core of the planet must precede the traction effect. The most significant 90-degree angle between planets was between Jupiter and Venus. Not only are they stronger in terms of mass and distance but they both probably have strong radiation as already mentioned. Mars also added to this in relating to Jupiter, as well as Uranus, and Neptune at 90 degrees to Saturn. Therefore, it appears that there was extra radiation heat input to the core. Here again records would probably clarify the situation.

Apart from the fifty earthquakes originally analysed, and the second fifty carried out as a check, I have constructed hundreds of these diagrams and repeatedly observed that planetary alignments at 90 degrees apart occur at such times. As well as this, the position of the tidal pressures appears to be significant. As stated elsewhere, this is different for islands than for larger landmasses. Even so, further analysis may explain other differences, as ocean tide pressures vary significantly. Either way, the cause appears to be from the external bodies in one way or another.

In summary, the indication is that earthquakes do not happen just because of internal processes but also from external processes. Figure 11 shows the six main tectonic plates. From it, we can see that the plates do not just drift; they move because of the rotational movement of the earth to the east, and because of tidal forces. As the earth rotates to the east, the surface landmasses tend to drag and appear to move to the west. In addition, the tide presses against the east coast of the land-masses as it tries to stay beneath the moon that has raised the water. This also helps to cause horizontal compression and to move the plates. Furthermore, the rotational movement of the earth is a planetary motion. So overall, with the effects on the core, we can risk saying that earthquakes do relate to the movements of celestial bodies.

The Latitude Effect with Earthquakes

The previous diagrams showed the position of the earthquakes in terms of longitude of the earth and the celestial longitude of the planets. However, there was no allowance for the inclination of the earth, as corrections for this are confusing on such a small plan type of diagram. The inclination of the earth, of approximately 23.5 degrees would make some difference according to the direction of the external traction. Added to that the orbital position of the moon varies as much as 5 degrees north or south. This orbit is not on the plane of the ecliptic, like the planets, but is on a shifting inclined orbit. Consequently, the specific line of traction from the moon would vary in relation to the inclination of the earth. This might therefore alter the line of force in relation to a fault, so that a similar diagram would not indicate a similar result. The simplest example of this is in relation to the moon and glacier movement. In Figure 12, the lower diagram shows a plan view with the North Pole inclined towards the external body. This is the moon and the traction field is shown as being almost over one half of the earth and the North Pole is well within that field, but the South Pole is not. In the upper diagram, the moon is at zero degrees celestial latitude but it can vary up and down from that. The traction field of the moon may therefore overlap the North Pole. Yet, at times it would not overlap. The long periods of daylight at the poles demonstrate this because the sun is then above the horizon. Conversely, the long periods of darkness are due to the sun being below the horizon. The point is that the indirect traction to the glaciers would not be a simple pull downwards, as the turning of the earth would probably create a downward corkscrew movement. If we apply this principle to fault lines, we can see that there is a shifting external traction all the time. That is to say, the external traction may not be pulling against a fault in the same line of force at any two different times. The same applies to the planets and the relationship varies because the earth is orbiting round the sun and its angle of inclination to the sun and to the planets will vary because of that. The planets only vary in celestial latitude one or two degrees and the sun is always at zero degrees. The solar plane is the plane of the ecliptic from which we take the other latitudinal measurements. The angular relationship of the earth continually changes, as seen in the seasons, and the angular relationship of the external bodies to the faults is always changing. What this indicates is that at times one fault could be disturbed and yet at another time a different one moves.

As an example of how this operates two more diagrams for the San Francisco earthquake are included. Figure 13a shows the latitude position as seen from the side and Figure 13b shows the latitude position as a plan view. The affected area is central to show the limits of the external traction. The inclination of the earth is included but the latitude of the planets and of the moon is not. If we compare these diagrams with a world globe in the correct position, the lines of force would be in relation to the ocean and the land.

Two further diagrams are also included for the Iran earthquake. Figure 13c shows the side view of the limits of the external traction. Figure 13d shows the limits of the traction from a plan view.

Analysis of Ocean Tides and Planetary Traction

In the original fifty diagrams used for analysis, the sun and the moon showed up as being most effective. They were the most critically placed and indicated that ocean and earth tide effects were the most likely trigger effect. In following checks, there were times when aftershocks matched the exact rising of the moon on the stressed area. Once an earthquake has occurred it is a simple matter of plotting the movement of the moon on the earthquake diagram. However the moon alone is almost ineffective. It is always in association with other forces, mainly the sun, but often planets also. Seismographic records studied to check this showed that the moon had little effect unless the fault had already moved. The lunar traction usually needed extra support from other forces. There was one isolated example that indicated this, It showed a very small variation on the graph. Someone commented that a sub-sidiary observatory had recorded a similar slight variation four minutes earlier. This was significant, and I checked the exact position of the other observatory and worked it out. In angular measurement, *it was exactly one-degree further east, and the earth rotates one degree in four minutes.* I had already checked the astronomical tables. Venus and the moon were both rising at the same moment. It indicated that the more eastern observatory was first disturbed as the bodies rose, and the western one four minute later, because of the distance.

In the table below, the New Moon is more prominent than Full Moon, probably because at a New Moon the sun and the moon are on the same side of the earth. They are together, in line, whereas at Full Moon they are on opposite sides of the earth. This indicates that in conjunction the combined traction is more effective. This is the principle of considering the planets. Their combined effects as a group will add to the gravitational effect of the moon or the sun. My conclusion from this, supported by the general studies, is that ocean tides alone are not the actual cause. In any case, if they were, there would be seismic disturbances matching almost every tide. Although they are very important, any one phenomenon is not effective on its own. It is a combination of effects. This starts with two planets affecting the core and finally a group of external bodies acting as the trigger. The ocean tide forces are always there but higher tides, caused by more traction, and the directional thrust of the tides, are variables that may be calculable.

The first line of the table shows that 22% of the fifty diagrams had the sun and the moon together in an effective traction position. This means that as a broad rule the impact would occur as the sun and the moon were rising or setting. In contrast, the Full Moon time only showed 14%. The two bodies together appear more effective. The Neap Tide periods showed up more frequently ie. 30%. A Neap Tide is at a time when the moon is 90 degrees apart from the sun and the tide is from both of them. The tidal water mound is still more towards the moon. It is not such a high tide as the Spring Tide because the water is more spread out. However, there are two Neap Tides in each moon cycle, at the first and third quarter, hence the higher figure.

34% of the fifty cases had planets exclusively applying to the point of disturbance. Overall, the sun and moon were twice as prominent as planets. Yet, in almost every case some planets were in a position to add to the effects of the sun and the moon.

TABLE 8–EARTHQUAKES AND MOON PHASES

Tide	Moon Phase	Earthquakes (50)	
Spring	New or near	11	22%
Spring	Full or near	7	14%
Neap	Quarter or near	15	30%
	Total	33	66%
	Planets only	17	34%

The Relative Mass of Planets and Seismic Activity

TABLE 9–RELATIVE MASS OF PLANETS

Planet	Orbital Period Days/years	Sun Distance Millions Kilometres	Distance to Sun (units)	Mean Nearest Earth Millions Kilometres	Nearest to Earth in units	Relative Mass
Earth	1 year	149.6	1.0	----	-	1.0
Mercu	88 days	57.91	0.39	91.69	0.61	0.06
Venus	225 days	108.2	0.72	41.39	0.28	0.82
Mars	1.88	227.9	1.52	78.34	0.52	0.11
Jupiter	11.86	778.3	5.20	628.74	4.20	318.0
Saturn	29.45	1427.	9.54	1277.41	8.54	85.14
Uranus	84.01	2869.	19.19	2720.0	8.19	14.52
Neptu	164.8	4496.	30.07	4347.1	29.07	17.46

One unit is one astronomical unit. This is the mean distance of the earth from the sun, which is 150 million kilometres (93 million miles). The relative mass of the sun is given as 332 958. The mean distance of the moon from the earth is 0.3844 million kilometres. Its relative mass is 0.0123. The moon has the greatest effect because it is much nearer. All celestial bodies are on elliptical orbits, therefore all the figures are normalised mean distances but there are times when planets are nearer and also moving at a different speed.

The table of relative mass makes it difficult to acknowledge any likely effect from planets. The bare figures suggest that any effect would be minimal. However, the effect is relative to the angle of force. We will discuss this further on. The mathematical calculation for tidal effects is not quite the same as the effect of gravity with two free bodies. In general, we multiply the masses of two suspended masses. We then divide that result by the distance squared but for tides, the distance is cubed. The comparative result is such a minuscule figure that it is even more difficult to consider that there could be an effect. Furthermore, astrophysicists consider that

Mercury would not be at all effective. It is only 0.06 the mass of that of the earth, which is counted as 1. In addition, because Mercury is so near to the sun it was difficult to see how any effect could be separate. Nevertheless, the diagrams indicated that overall that there might be some effect. In considering this problem, it seemed that the clue must be in terms of indirect traction because classical gravitation models are always in terms of direct traction. That is to say, the measurements are in terms of one total mass related to another in a direct line of force. With indirect traction, it is not a direct line of force to the mass as a whole but in effect, the traction is bypassing the gravitational pull of the main mass of the earth. In using Mercury as a standard of traction, we therefore only have to consider Mercury against one tectonic plate in a specific position and not against the whole earth.

We can demonstrate this by considering one plate. The plates ride on the upper mantle or asthenosphere at an average depth of about 100 kilometres. It actually varies between around 50 to 240 kilometres but for purposes of demonstration, a round figure is adequate. There are six main plates and if we take one sixth of the surface of the earth and use the given depth of 100 kilometres, we only get a volume of approximately one hundredth the total volume of the earth. In terms of mass, this will be less because the earth is denser at the core. In these terms, Mercury then has a ratio to one plate at an advantage of 0.06 to 0.01. In a direct position, Mercury would be competing with the gravitational force of the whole mass of the earth. However, in a force-line of indirect traction, the pull is not only bypassing the main mass but it is pulling the plate sideways and not outwards. This is the key to how a relatively small mass, in terms of traction, can affect a fault. This shows with a simple experiment using a weight and a spring balance. If a 5-kilogram weight suspends from the spring balance, it will show the full 5 kilograms because the gravity of the earth is pulling against the spring measure. If the weight is dragged along the ground, sideways, the weight can move with much less than a 5 kilogram reading on the measuring scale. On a smooth surface, it is around one quarter of the lifting measure, and on a rough surface, it is more. The diagram in Figure 14 explains the principle.

The tectonic plates are of course sandwiched between other plates, yet the other simple model, of a magnet pulling against curved strips of metal showed that they all moved. In other words, the direct traction is pulling the nearer ones out of the way and the ones at the tangent position *slide sideways*. The raising of the earth at the point of direct traction helps in this. The point is that with extra traction there is enough movement to help the plates at the low tide position to move sideways. So on the entry side there is jerk forwards as it enters the field and on the other side, there is a sudden jerk as it escapes the field. The plates would slide more easily once the mantle became softer and adjacent plates were slightly out of the way. In this way, a small mass might be "the last straw" that triggers an earthquake. Computerised models could demonstrate this, and the other points.

As a further example of how the sideways movement is easier, we can consider moving the hands of a large clock. It is easy to turn the long hand in a circular motion with only a slight pressure against the tip. However, it is virtually impossible to pull the clock hand outwards in the direction in which is pointing. The direct pull is analogous to direct traction, and has little or no effect. Yet, a slight sideways pressure, which is analogous to indirect traction, moves the hand quite easily. Of course, clocks are specifically designed to operate in this manner, but the analogy does help to explain why a distant or small planet may have some effect.

CHAPTER 4

FORESEEING EARTHQUAKES

Aids to Early Warnings of Earthquakes

Geophysicists have various highly specialised methods of checking for seismic activity. They include such methods as measuring magnetic variations and electrical energy, recording any increase of charged particles from the earth and testing for the emission of radon gas near faults. With all these sophisticated techniques, specialists are still wary of predicting earthquakes and some have openly stated that it is impossible. In offering my own findings I wish to state that I prefer to say that indications of likely seismic activity could be more accurately foreseen with knowledge of these external effects.

Before I offer my conclusions, I would like to make two observations that may be relevant. In literature on detecting seismic activity, there was a reference to the use of a gravity meter. As I understand it, this is for measuring variations in the gravity. The reference alluded to the frustrating operation of the meter at the end of the day. The "end of the day" appears as significant and my comment is that the variations in the readings may perhaps have been caused by the sun, and other bodies, setting at that point. This would also be the case as the sun and other bodies rise, that is, as a group. If this were correct, it would indicate a different gravitational effect at the low tide position, ie. where the external bodies were either rising or setting.

The other observation concerns magnetic effects. A field research officer who worked on oil exploration helped with this. His duties included making magnetometer investigations. I asked him if he could send me a printout of any readings where there were inexplicable variations, especially at dawn or sunset. I was only able to obtain these readings for a short time but the results were interesting. The time was on the readout so I was able to make diagrams for those times. The few I examined showed a very slight "blip" as a strong group of bodies, including the sun, rose over the horizon. I am not in a position to make a firm statement about this as it is a specialised area, but observation of such times may perhaps be helpful.

From these different observations, it seems there could perhaps be an extra means of foreseeing seismic activity. My firm conclusion is that there could be advanced indications and warning signs that seismic activity is likely. This could help specialists at ground level obtain an extra understanding of the developing disturbance. A list of the possible aids is below.

Astronomical aids to forecasting earthquakes

- Observing the development of a 90-degree angle between any two planets.
- Observing any build up of electromagnetic variations as the planetary alignment changes.
- Observing ion count or magnetic variations at the times when planets are rising or setting.
- Observing groupings of planets.
- Observing seismic activity to the East relative to any applying traction from external bodies.
- There may be a Doppler energy effect when the earth is orbiting *towards* a planet. This may then modify the effect.

Apart from the specific astronomical factors relating to the seismic area the external traction is all around the earth, and earthquake areas to the east would be in the traction field earlier. This became obvious by the incident already mentioned of a slight tremor further east a few minutes earlier. Because of this, I checked official lists of earthquakes to see if there were tremors further east when any particular fault was disturbed. There were many disturbances where this was the case. The distances between the seismic areas converted into times that matched the transit of the external bodies. However, there were some variations and my view is that any useful forecasting is only likely to be from specialists who are observing one particular fault.

Modern research methods mainly relate to the local fault, near the geophysics observatory. The methods are similar, with some variations. The present trend, in the 1990's, is towards measuring the changes of electric potential in the rocks under stress. Yet, this is still only dealing with one part of the chain of effects. My standpoint is that the process starts from the planets. Awareness of planetary causes would give ground level researchers an extra pointer in applying their own expertise. As I see it the refinement of predictions will be with localised geophysical knowledge and knowledge of the planetary effect could help in that refinement.

Personal Observations in an Earthquake Experience

In studying this situation, I used a diagram for any particular day like a clock. The sun is the position of noon, local time. The earth rotates 15 degrees in one hour. This can be marked off on the circle so that in local time the rising or setting of any external body is clear. It is not completely accurate as there is no allowance for the inclination of the earth, but it is a useful guide. When I was doing this main study, I lived in an area where there was an earthquake to the north. The shock and after shocks were easily felt. In this particular case the moon showed as being particularly effective as it aligned with grouped planets. When the first shock struck we all ran outside and watched the house shaking. It was a wooden house, on high stumps and was visibly shaking. Then it was all over. I rushed inside and made a diagram for the time of the shock. It was perfect. The after shocks lasted for some days and I plotted the position of the moon forward every day. I made two interesting observations. One was on the next day when I went outside to watch the moon rise and check the time. There was a slight tremor as the moon rose. In the following days, I plotted times to anticipate tremors and for three days could feel them just where the diagram indicated extra stress. As this is the only time in which I had actual experience of tremors, I have no way of confirming how often such predictions could be useful. To me it confirmed my conclusions, and overall I feel confident that some forecasting is possible.

In using these principles to assess any likely effects, the actual distance of the planets is important. For example, Jupiter is 5.2 astronomical units *from the sun,* but when the sun and Jupiter are in line, it is 6.2 units *from the earth.* This is because it is then on the far side of the sun. Jupiter is at its closest to the earth when the sun is on the opposite side from the earth. It is then only 4.2 units from the earth. The moon and Jupiter can then make a stronger traction force. A similar problem exists at times with Venus as it can be on the earth side or on the far side of the sun. My statements that the sun and the moon, as well as Venus and Jupiter are the strongest traction forces, does not mean that they always have the same effect. In comparison Saturn, Jupiter, Mars and the moon, on the opposite side, away from the sun, could have a stronger effect as Saturn and Jupiter are then at their nearest distance to the earth. Questions as to why there is an effect one time and not another may relate to these points. Nevertheless, it is noticeable that there is often an effect when Saturn, Jupiter, Venus, the moon and the sun are closely in line.

Missing Factors

The diagrams in Figure 22 summarise the main planetary positions that appear to relate to the development of seismic activity. Although we will consider them again later, it may be of interest to mention a few details here. Scientists have repeatedly referred to some of these points, particularly the suggestion that most planets on the same side of the earth might cause extra gravitational stress. In addition, they had examined tidal patterns. As mentioned elsewhere these related to the times of high tides. There were many references to high tides but no indication of what happened at low tide. This led to consideration of the low tide position. The investigation further led to the conclusion that there was stress at the low tide position as the fault moved into and out of the combined gravitational field of the external bodies. An analysis of many major earthquakes indicated that this appeared to be the case. However, in attempting to use this as a means of foreseeing seismic activity world-wide, the process was not always reliable. The problem was that there were many such close groupings with most planets on the same side of the earth and yet there was no major seismic activity.

From this it seemed that there was another factor. What was missing was the essential situation of at least two planets being at a ninety-degree alignment as a means of stimulating the core of the earth. The interesting point was that all the diagrammatic information was in the study of sunspot cycles. Nevertheless, they did not seem related and the two studies were virtually separate. However, when the processes appeared to be similar, the missing factors began to show. The point at issue here is that if the core is not stimulated the magma will not be softer and the plates will not move easily. At such times, even a strong gravitational group will not have much effect. The effect on the core from the external bodies appears to be the initial necessity for the other factors to apply.

In drawing attention to this, it seems that not all the planets have the same effect on the sun. From this, we can perhaps conclude that they would equally not have the same effect on the earth. In other words, different combinations may be more effective and some combinations may affect the core more than others do. Apart from that, there are probably still other factors and the suggestion of five contributory causes is probably restrictive. The further consideration of solar storms and ionospheric disturbances gives an indication that there is a wider process.

The Doppler Effect

The suggestion that there may be a Doppler effect rests on the observation of the position of the major planets at the times of seismic activity. Jupiter and Saturn are apparently the most likely for such an effect, as they were noticeably more prominent in the solar cycles. However, we have two possibilities. One is in terms of a gravitational effect and the other is in terms of electromagnetic forces. Either way the principle is broadly the same. As already pointed out the major planets can be on the far side of the sun or on the earth side of the sun. The difference in astronomical distance is two astronomical units. This is because the earth is orbiting the sun. On one side it is nearer to the planet and on the opposite side it is further away as the sun is in between the earth and the planet. The first point therefore is that at its nearest a major planet would have a stronger gravitational effect and hypothetically could also then have a stronger radiation effect from the electromagnetic waves. Astronomers know that radio signals are fainter with increased distance. Their experience with probes to outer space has demonstrated this. Proximity should therefore strengthen the effect. In this process, all the planets are moving and the earth is orbiting round the sun. As it moves round, there would be a point where the earth was receiving a stronger signal. Radio astronomers may know this and it would appear to be the point where the planet begins to go retrograde. This is an apparent motion caused by the earth moving round to the same side of the sun as the planet. In effect it is "turning the corner" to start moving towards the planet. Hypothetically, there are then two critical positions. One is where the earth is receiving the strongest signal in its more direct movement towards the planet. The other is when the earth is at its nearest to the planet.

These factors may be part of the missing equations that might help in forecasting seismic activity. Forecasting the weather relies on information from many sources and computer models indicate the most likely weather pattern. Theoretically, adequate information could help to indicate the most likely seismic effect. There are undoubtedly other unknown factors. Understanding these usually starts from observation and then speculation about the significance, followed by testing. In this, it is observable that the relative position of the earth is significant. Whether there is a Doppler effect with gravity as well as electromagnetic waves is another area where expert knowledge will provide some answers. References to the Doppler effect describe details about light and sound, and theoretically, they could apply to other forces and energies.

CHAPTER 5

VOLCANOES

External Forces with Volcanoes and Mountain Building

Volcanoes relate to mountain building and the joint process relates to ocean tidal forces. A 'rim of fire' surrounds the Pacific Ocean, because there are so many volcanoes on the shores of the Pacific Ocean. Activity within the earth is the cause of these volcanoes bursting into life from time to time. The conclusion offered here is that inner activity is by stimulation from the planets, as already explained. However, the Pacific Ocean is the largest mass of free water and therefore capable of creating stronger tide forces by the piling up of the water. These tidal forces help to cause horizontal compression. This adds to the natural drift of the tectonic plates. The result is crumpling of the earth's crust and a volcanic outburst if the mantle and the magma are excessively heated and under pressure. The basic process is similar to that with earthquakes and it is interesting to consider how one effect causes an earthquake and the other a volcanic eruption. Of course, there are seismic disturbances when there is an eruption. From the study of many diagrams for volcanic eruptions, there were two main differences from earthquake diagrams. The first is that there were generally more than two planets at near 90 degrees apart. This suggests that the core may therefore receive more stimulation and the magma is hotter. It is a point worth examining. In the case of volcanic outbursts, the investigation also showed that the planetary alignments were generally closer. They were usually in the same general direction as the sun, so when the moon was also in that direction there was a much more definite ocean-tide effect. The height of ocean tides to the east, at the times of volcanic effects, would probably show higher than normal tides.

In general, all the bodies were on the same side of the earth, but seismic disturbances do also occur in such alignments. It is not therefore a clear rule but only a generalisation. The suggestion is that in this positioning there is a more definite "push-pull" effect on the affected area. This is when it moves into and out of the traction zone. The most stressed area is where the ocean tides are applying their force. It may then be some distance from the actual volcano. Therefore, after the first effect from two or more planets creating an energy stimulation of the core, there is the second effect. This is from the ocean tide compression, and the third is from planetary traction applying directly to the affected area. These effects would help to create the seismic shocks but not necessarily cause the eruption. The operative point here is that the low tide indirect traction helps to cause a sideways movement of the crust, but a high tide from direct traction causes the extra water that is striking the coast-line to the east.

Of course, this is not as simple as it sounds, as these are conclusions based on an ideal model. In real terms, ideal conditions never exist. As an example, the diagrams Figure 15 show the two ideal effects of a high tide and a low tide. In one diagram, the landmass is in the centre of the high tide mound of water. (Both views are looking at the South Pole so in this position the earth is rotating clockwise.) In the diagrams, the crust is in six equal sections. These indicate the main plates. If the traction forces are directly above the area, there could be pressure from the ocean on both sides but as the earth rotates, it would change. If the position in Figure 15 is the starting point, the land has to move against an ever-increasing pressure of water that is trying to stay beneath the moon and other external forces. Then from the direct-position it has to force its way through the high mound of water to escape the pressure. At the low tide point, the water is pulled away and the land is also strained by an earth tide but the earth itself keeps rotating. Consequently, there is not only the "push-pull" effect, but also a "bumping action" where the horizontal pressure travels to the volcanic area.

The combination of these two effects; the initial stimulation of the core and then an ocean-tide push and pressure against the land and on the shelf, affects the magma and the crust. Astronomically stronger and slower angular alignments and closer grouping of the external bodies create a more definite effect. Figures 16a-c demonstrate these principles.

Planetary Positions at Volcanic Eruptions

Figure 16a is for the highly dramatic eruption of Krakatoa (also known as Krakatao) in 1883. It literally blew the island apart and the dust cloud created fabulous sunsets right round the world for some time afterwards. Krakatoa is a tiny island in the Sunda Straits, between Sumatra and Java. In terms of possible tidal effects, it is in a critical position. The land of the Eurasian Continent blocks the high tide mound of water in the Pacific Ocean. The water can therefore only try to sweep westwards, to stay under the moon, by attempting to go round the large mainland. This is via the Bering Straits in the North Pacific, or round the Indonesian islands into the Indian Ocean. Kamchatka, near the Bering Straits, has many seismic shocks but much of the pressure is towards the south as the water can more easily go westwards past the islands. A global map clearly demonstrates how the pressures operate. The peninsula of Kamchatka runs approximately southwest, towards Japan, and with the islands of Japan forms the first obstacle to the water moving westwards. The mountain formation is noticeably along the spine of the islands, and is especially prominent along the Kamchatka Peninsula. These mountains are clearly the result of continual ocean tide pressures and the mountain formation in China is the product of horizontal compression from these tidal forces. What is more the continuation of these mountains indicates that the ocean tidal pressures are not only on nearby coastal areas, but can be transmitted long distances. Mountain formation may therefore be on either coast, or inland according to how the forces are able to exert adequate pressure. The proximity of faults relates to this process.

The diagram for Krakatoa is a typical volcanic pattern of alignment. In the diagram, Uranus and Neptune are not strong traction forces but they are in suitable alignments for affecting the core of the earth. Uranus was 83 degrees to Mars and 105 degrees to Saturn, and Neptune was 90 degrees to Venus. As volcanic pressures build up slowly, earlier alignments that might stimulate the core are important. Overall, it appears there was strong core stimulation. As already mentioned this may be the difference between a volcanic eruption and an earthquake. If the core is hotter because of more stimulation, the magma is more fluid. Comparison of ground heat in the different areas might indicate this. In the Figure 16a diagram, "C" indicates the main direct traction. The rotation is anti-clockwise, towards the east and "D" is the point of escape from the direct traction. "A" is a "quiet spot" with no planets applying to that area and "B" shows the re-entry into the traction field. During a one-week period, all the planets are fairly well in the same position. The sun moves one degree each day and the moon moves approximately 13 degrees, so in one week could move round one quarter of the globe. It would have been exactly in line with Saturn the day before the eruption, but pressures had probably been building up for some time. The other examples are much more recent and accurate records are probably available for a detailed check on any associated phenomena.

There are more volcanoes in the southeast pacific area than anywhere else. My conclusion is that this is because of the ocean tide pressures of the Pacific Ocean. They are unable to escape and pressure builds up on the crust in that area. Finally, there is one interesting coincidence with Krakatoa. The explosion was in 1883 and the sunspot maximum after the irregular activity shown in Figure 2a was in 1883. It seems that because of connections in solar and terrestrial phenomena unusual variations in sunspot activity would have unusual terrestrial effects.

Ocean Tides and Eruptions

The second diagram, Figure 16b, is for a Mt Etna eruption on August 5 1979. At the time of the explosion, it was just past the point of Neap Tide. The moon (8) was not quite risen. It would therefore have been a position of near low tide in terms of the lunar traction but the sun (1), Jupiter (6), Venus (3) (and Mercury) were applying direct traction. The Neap Tide position would be between the direct position of the sun, (1), and the direct position of the moon, (8). The eruption was just before noon, local time, when the sun is overhead. (This shows how local time is easy to apply). The moon was more than two hours from rising. The diagram in Figure 16c is also for Mt Etna and shows the situation for the earlier activity on July 16. It is interesting to plot the position of the moon in such sequences. The next diagram in Figure 17 shows the position of Mt Etna in relation to the Straits of Messina. Tidal variations are somewhat lower in the Mediterranean Sea, and as a point of interest, the Romans ran into difficulties in their first attempts to invade Britain, in 55 BC. The currents and tides in the English Channel were different from their local experience. This is because there is the larger mass of water in the North Sea trying to get round the British Isles. It demonstrates how and why tidal forces vary in different parts of the world. The diagram of the Straits of Messina shows the curve in the land that would help water to press against the coast and create horizontal compression.

The third example, shown in Figures 18a, 18b and 18c refers to Mt St Helens, in Washington State, USA. This 1980 eruption is of unique interest because it starts with a scattered diagram on March 20 for the first tremor. The media covered it in detail and later there was a report of all activity in technical geophysics journals. As it made world news, I was able to follow it day by day. The times were all local times in the journals, so I have used them. Four pairs of planets were at or near the 90-degree angle. They were Neptune/Saturn, Uranus/Mars, Uranus/Jupiter and Mars/Venus. As a point of comparison, Krakatoa had three pairs of planets at a suitable angle and Mt Etna had four pairs. This suggests that there probably is extra core stimulation for volcanic activity. The first tremor occurred just as Mars, Jupiter and Saturn were near rising, and the moon was overhead (with Venus). The moon indicates the approximate position of the high tide, but it is probably further east. This is because there is always some delay in the drag. Venus would be adding some traction, but the sun, approximately 60 degrees to the east would still be having a Neap Tide effect. It is almost the same type of effect shown in Figure 15. The next effect was of harmonic tremors, after the time of Full Moon. The sun had just set and the moon was rising (Figure 18b).

Figure 18c shows the diagram for the final explosion on 18 May. Note that the sun, Mercury, Venus and the moon were rising on the affected position. New Moon was on the 14 May. The traction was at the low tide position, and all the planets are on one side of the earth, as with Krakatoa and Mt Etna.

These three diagrams show how difficult it is to focus on one definite pattern. Earthquakes are much more specific in time and position. The principle is the same, but volcanic activity is more drawn out as a related aspect of mountain building.

Figure 17 is for the Afar Triangle relative to the Gulf of Aden. A field expedition with the vulcanologist, Haroun Tazieff, went to study that area in 1967. According to the report in the *Scientific American* (1970), the layers of rocks were tilted in the opposite direction to that expected. The report said that the area had also had at one time been under water. From this we can apply the tidal principle and see that tidal waters trying to move into the Red Sea would cause pressure at that point. The rocks would therefore incline towards the west. It is a good example of how tidal pressures help to create mountain building. The tidal water would not be able to escape from the Red Sea and, just to the east, they would not be able to escape from the Persian Gulf. It is interesting to compare the Afar Triangle with the entrance to the Persian Gulf. The mountains are mainly along the northern side of the Gulf of Oman. These are an extension of the mountains from the northeast. However, there are some mountains on the lower east of the Gulf of Oman. These would appear to be similar in creation to those in the Afar Triangle. Either way it shows that tidal forces help in all these processes. The one difference that appears to stand out in comparison is that mountain building is not necessarily dependent on internal stimulation of the core of the earth whereas earthquakes are. Volcanoes are therefore a dual, or combined process. The mountains in the Afar Triangle are volcanic and some of the volcanoes are continually active. It would be interesting to compare their recurring activity with tide levels and the grouping of any planets that help to create a higher tide in that area of the Red Sea. The "parting of the waters" that helped Moses escape from the pursuing Egyptians could have been an exceptionally lower tide and the incoming waters that engulfed the Egyptians would be the corresponding higher high tide. Planetary alignments for that time would probably match the phenomena, but it is unlikely that we know the exact time.

Vulcanologists are particularly interested in foreseeing volcanic activity as a means of warning people. With the aid of astronomical knowledge, it might be possible to foresee the likely critical climax, which could then add to other methods.

For example, one volcanic eruption that followed a pattern occurred in the Philippines in 1991. The main activity started earlier in the year and the eventual climax was on the 15th/16th of June. The alignments were interesting. The major planets Jupiter and Saturn were almost opposite each other, with Uranus and Neptune close together on the same side as Saturn. Venus and Mars were near the same longitude as Jupiter. At first glance, it appears like a "weak" pattern. However as the activity had started some time earlier it would have been possible to foresee likely critical points in time. To start with, Jupiter and Saturn were almost exactly opposite each other on March 22 and Venus was exactly at 90 degrees to both of them at that time, so here are two interesting 90-degree angles. In addition, two weeks earlier, Venus was at 90 degrees to Uranus and Neptune. All these would stimulate the core. Next, there is a modest cross-traction of Venus and Mars against Saturn with the moon adding to their traction. There was a New Moon on June 12, followed by the moon in exact alignment with Venus, Mars and Jupiter on the 15th/16th. If we ignore Mars, we have the four strongest traction effects close together, ie. the sun, the moon, Venus and Jupiter. Comparison with tide levels for that time may indicate the extra pressure. The important point is that earlier alignments were contributory to the outcome. Once local activity had started observation of alignments at recurring activity could give extra indications and perhaps help with warnings. With earthquakes, there are sometimes no earlier shocks and no warning, but with volcanic eruptions, there is usually an increase in seismic activity that could show against tides and planetary traction positions.

Finally, on the subject of planetary stimulation it is likely that at times the operative planets such as Jupiter, or Venus, might themselves have recently received an effect by critical alignments. In such a case, they might then have a stronger radiation than usual. This would show by comparison of radio noise with alignments to it. My view is that the earth, and the sun, are not isolated self-contained objects, but are part of an integrated electromagnetic radiation system that affects all objects within its field. It would be restrictive to assume that only the earth receives effects in such a manner. The planets are all affecting each other and in their role as senders and receivers of the energy, they might therefore be more effective at some times than at others.

Nevertheless, there are some odd exceptions to the general "rules". For example, the earthquake in Quetta on 30 May 1935 occurred at 3.00 a.m. local time, and the diagram was a very scattered pattern of alignments. On 16 August 1959 at midnight, there was an earthquake in the Yellowstone National Park, also with a scattered pattern. The interesting point with these earthquakes is that both reported unusual prior phenomena. In Quetta the birds flew away, the animals were restless on the previous evening, and unusual atmospheric electrical phenomena occurred. In Yellowstone, in the National Park, birds flew away at noon, and the shock occurred at midnight, twelve hours later. The diagram showed that at noon, Jupiter was just rising on that area, and at midnight, Jupiter was just setting. For Quetta, Jupiter rose at approximately 6.00 PM. This was when the animal restlessness and phenomena began. The earthquake occurred when Jupiter was near setting. This suggests related phenomena, perhaps in the area of electromagnetic effects, and not gravitational forces on their own. In any case, there may also be an extra input of energy from a solar flare. I have checked the incidence of many flares against earthquakes, and it seems that the process may relate to seismic disturbance. Further details are in Chapters 11 and 13.

The basic principle, which the further comparisons appear to suggest, is that there are internal tidal effects with volcanoes. The initial pressures are apparently similar to the seismic effect, but when the external bodies are overhead, there would be a high tide effect in the magma. This would help to cause an eruption. Scientists have already considered a possible tidal effect in the magma when it is very fluid, due to increased heat. From this viewpoint it is likely that observations in terms of the principles suggested here could help to clarify the process.

Finally, scientists have determined that the volcano of Krakatoa erupted violently in 535 AD. In that period, the major planets were in a critical arrangement. Uranus and Neptune were almost opposite, which generally appears to coincide with low internal activity as demonstrated by seismic activity. However, it appears that such periods can sometimes be one of unusual or "irregular" activity, according to how the other planets match into the alignments. The point of interest on this is that on January 1st 536 the major planets were in positions that hypothetically would help to stimulate internal activity in the earth. The geocentric alignments in coordinates of Right Ascension, in hours and minutes, were as follows. Neptune 21:43, Uranus 9:38, Saturn 10:12, and Jupiter 15:04.

This means that Neptune and Uranus were 177 degrees apart. Saturn had recently aligned with Uranus and would add to any effect. In addition, Jupiter was 87 degrees from Uranus, 83 degrees from Saturn and exactly 90 degrees from Neptune. In terms of the overall analysis, this would therefore not be a quiet period but would be unusually active. The further explanations will help to make this clearer, and the actual heliocentric positions would modify the effect via the sunspot process.

CHAPTER 6

GRAVITY AND ELECTROMAGNETIC FORCES

Critical Planetary Grouping

The original study discussed in Chapter 1, considered the specific grouping of planets as an additional gravitational force that would help to trigger earthquakes. In this, I worked on the conventional principle of different masses affecting each other. As a working hypothesis, it is the basis in all the previous investigations. However, there has been a recurring possibility that there might be an additional explanation of the phenomena. For example in considering sunspots it was posited that the sunspots are disturbed because they have rotated into a planetary gravitational field that disturbed the sunspots and caused a flare. Yet, we could perhaps equally say that the sunspot is disturbed because it has rotated into a planetary electromagnetic field. This would not necessarily supercede the gravitational effect but might operate with it. In other words, hypothetically there could be an integrated and interlocking field system. In this, the gravitational forces and low intensity electromagnetic forces would operate together. This system, if it exists, would operate with sunspots and earthquakes. The question is, are there any indications that this might be feasible? Recent research by others indicates that there might be some subtle effects.

One report, obtained from the Internet, referred to a unique prediction of seismic activity by Dr Elizabeth Rauscher and W.L. Van Bisc, dated January 8 1994. They predicted that quakes would occur in or near the Los Angeles area within the next 30 days. They based this on the increase of superconductivity in the rocks. On January 17 the Northridge California earthquake struck, with a magnitude of over 6.0 on the Richter scale. The report went on to say that there were unusual low frequency electrical surges *beginning two weeks before the quake*. I therefore looked at the planetary alignments for that time. Two weeks before the shock would be January 3. The planets were in a close group on January 17, but were almost in the same direction on January 3. At that time Mars, Venus, Mercury and the sun were almost exactly in line. To the nearest degree, counting anti-clockwise from zero, the geocentric positions were Venus 279, Mars 281, Mercury 281, and the sun 282, all within an arc of 3 degrees.

Unique Electrical Effects

This alignment coincided with the beginning of the electrical effects. Could there be a connection by a different electromagnetic effect from planets, especially Mars and Venus? The interesting point about the two planets is that they are perhaps both positive. In the sequence of planets, discussed in Chapter 1, with the earth as negative, both Venus and Mars appear as positive. In my original study of fifty earthquakes, I had noticed that Mars and Venus were often close together. Out of the fifty major earthquakes, twenty-four showed Mars and Venus within 60 degrees of each other, and seven were over 60 degrees and less than 90 degrees apart. The others were outside those limits. On the face of it, such angular proximity suggests a combined gravitational effect and this was how I had applied it. However, in such cases, Mars is on the far side of the sun and its gravitational effect would be somewhat slight. There might perhaps be another effect, such as an electromagnetic connection.

Venus could be on either side of the sun in such an alignment and the sun would make it difficult to separate any electromagnetic effects. Nevertheless, it is uniquely coincidental that the ground electrical effects started at that time. It suggests that the combined "positive nature" from those two planets might perhaps have an effect on the "negative earth".

The suggestion from the positioning of these two planets is that there may be two effects. One is on the earth and the other is on the sun. In the first supposition, there would be two positive planets affecting a negative earth. In the second suggestion, there are two positive planets affecting the sun that may also be positive in these terms. It is an area for specialist research and we cannot here assume specific effects. Nevertheless, scientists know that the sun and the moon have a tidal effect on the ionosphere. This is often in association with solar disturbances and magnetic storms but the literature does not refer to tidal effects from the planets. It may therefore be that the electrical effects relate to ionosphere disturbance by the planets and not just to ground effects. Furthermore, it may even be that the planets in a ninety-degree alignment have a direct effect on the ionosphere itself. Since the ionosphere consists of charged particles in a radioactive state, the effect of the electromagnetic wave radiation may perhaps be somewhat similar to the effect in the core of the earth. Of course, it will not be the same because the core of the earth has a condensed mineral structure, which is highly compressed and intensely hot. At this stage, these conjectures may appear as supposition but the coincidence of the different alignments may relate to more than one effect.

As an indication of the repetition of alignments for Venus and Mars, a brief table is below. Mars and the sun are exactly in line over a period spanning just over two years. Venus, as an inner planet, is anywhere on its orbit at that time. It may be on the far side of the sun, on the earth side of the sun or at an angular distance to one side or the other. Mars and Venus can both make variations with the sun. The second list gives the dates when they are exactly in line as viewed from the earth and the angular distance shows how far they are in advance of or behind the position of the sun. Their positions in heliocentric coordinates may perhaps also be important if there is indeed an electromagnetic effect from them. Although it is only conjecture, it is possible that our two closest neighbours could be important in a unique effect. This may show up better in computer analysis.

TABLE 10–FREQUENCY OF MARS/VENUS ALIGNMENTS (GEO-CENTRIC)

Mars/Sun Alignment		*Mars/Venus Exact Alignment*	
November 6th	1991	June 26th	1991
December 25th	1993	February 18th	1992
March 4th	1996	January 6th	1994
May 10th	1998	November 22nd	1995
July 1st	2000	June 30th and September 3rd	1996
		October 25th	1997
		August 5th	1998
		June 23rd	2000

The Northridge Earthquake

The close positioning of Venus and Mars on January 3, just before the Northridge series of earthquakes that started on the 17th, offered an opportunity to examine the reports in more detail. The Internet based *NCEDC Earthquake Catalog and Phase Data* provided a list of 163 quakes of over 3.0 magnitude for January 17 1994. In that list, 145 were for that area and only 18 were elsewhere. Of these 12 were before 12:30. The first shock of 6.6 magnitude was at 12:30 GMT and the last (on that date) was at 23:33 GMT. In that time, there were 19 above Magnitude 4 on the Richter scale.

Los Angeles is 33.50 latitude north, and 118.22 longitude west. The shocks were at latitude 34.30 (+/-) and longitude 118.50 (+/-), but only varied by a fraction of a degree. For discussing the earthquake we can therefore take 118 degrees west as the essential point. This is the position of Northridge in the diagrams. A plan diagram as shown only gives a close approximate position because there is no allowance for the inclination of the earth. The error varies according to the relative position of the area to the inclination. A diagram in the form of a technical drawing is in Figure 19. This shows the traction was not exactly on the affected area, but was on half the landmass, as the eastern coast is some distance away. The diagrams are only for discussing the principles. The ideal practical method is with three dimensional computer graphics.

A list of times of disturbances above magnitude 4.0 at Northridge are given below as a means of plotting their position relative to the alignments of the planets. The times show the position at different time. The times are UT/GMT but if the position of the sun is 12, the time matches local time anywhere in the world. For students in a seismic area it is a quick, useful and interesting method of observation. One hour is 15 degrees and 1 degree is four minutes of time. UT/GMT is also useful for plotting earthquakes on any one day. With a variety of earthquakes, it is evident that not all areas relate to the same alignments. Some faults escape disturbance when others do not. It therefore appears unlikely that astronomical methods alone could offer specific predictions.

TABLE 11–NORTHRIDGE EARTHQUAKES TIMES AND EFFECTS

Time (GMT)	*Magnitude 4+*	*Comments*	
12:30	6.6	These details are for January 17th	
12:31	5.8	1994. In the first half-hour, there were 25 shocks of magnitude 3 or more. The progress is below.	
12:34	4.4	*Time*	*Number of shocks*
12:39	4.8	12:30 to 13:00	25
12:40:09	4.8	13:00 to 14:00	24
12:40:36	5.1	14:00 to 15:00	14
12:55	4.0	15:00 to 16:00	11
13:06	4.6	16:00 to 17:00	9
13:26	4.6	17:00 to 18:00	3
13:56	4.4	18:00 to 19:00	5
14:14	4.4	19:00 to 20:00	11
15:07:17	4.1	20:00 to 21:00	14
15:07:25	4.1	21:00 to 22:00	2
15:54	4.8	22:00 to 23:00	11
17:56	4.6	23:00 to 24:00	8
19:43	4.0	Total	145
20:46	4.8	Above Magnitude 4	19
22:31	4.1		
23:33	5.5		

The Northridge Diagram

The diagram in Figure 20 shows which planets were rising or setting at any particular time, and we can evaluate where the maximum traction might be at any point. The arrow for 118 degrees west shows the position of Northridge at the time of the first shock. Previous observations show the sun, Jupiter, Venus and the moon as having the strongest gravitational effect. Any two of these in a group or only as a pair appear to be effective, especially with the planets at their nearest position. The main group included the sun and Venus. The earth was at its nearest to the sun in early January but Venus was not at its closest. We can see that the main group would be rising about one hour after the time of the first shock. This would put the main ocean tide effect in the East Atlantic Ocean and be pulling the waters out of the Gulf of Mexico. In other words, the tides would be relatively low in the Gulf. This means there would be less water on the shelf and it would be easier for it to spring in that area. With less tidal force pushing against the east coast, the land could move eastwards. This implies that at such a time the surface of the earth stresses in a different manner from another direction. This becomes clearer if we compare this diagram with Figure 8, for San Francisco and Figure 10, for Mexico City. For San Francisco, the traction was pulling to the east, as with Northridge. For Mexico City, the traction was pulling west. The diagrams in Figures 8, 9 and 10 show the three types of traction effects: (a) pulling to the west, (b) cross-traction and (c) pulling to the east. The ocean tides vary with the position of the moon, as previously explained. My own view is that forecasting and prediction would depend on understanding these three types of applied force and knowledge of tides in local indications.

The Earthquakes and the Planets

In the diagram, (Figure 20) the number of tremors for each hour is underlined. As already explained the other numbers indicate the

Universal Time (Greenwich Mean Time) in relation to 118 degrees west (not local time). Zero-degree longitude is in line with the sun at 12 noon. Plotting from that point would give the times when zero-degree longitude was at any point on the diagram. Here UT or GMT is relative to 118 degrees west as a means of using the diagram like a clock. In the first one and half-hour period, there were 49 shocks over magnitude 3. Ten of those shocks were over magnitude 4. The number of shocks then decreased as the area escaped from the traction. Between 17:00 and 18:00 hours, there were three shocks. At that point, the area escaped from the indirect pull of Saturn and then moved to a point where the moon was near rising. As the moon came near to rising the number of shocks increased. At that point, where the moon was rising there was an increase to eleven shocks. It was also a position with cross traction as Jupiter was nearly setting. When Jupiter was exactly setting, there was an increase to fourteen shocks then a drop to two shocks. My evaluation is that at that point, there was a point with no specific traction and there would be a slight release effect, or "elastic rebound". The area then started moving towards the setting group.

The ocean tide position could also be important. It is mathematically possible to calculate exactly where the actual high tide mound of water would be, but for the purpose of this example, we can estimate it. The sun and the moon were just over 70 degrees apart. The Neap Tide occurs when they are 90 degrees apart. The high point is always nearer to the moon and my conclusion is that the high tide water was now sweeping into the Gulf of Mexico and reversing the original tidal pressure stress.

The details for the Mexico City quake and the diagrams in Figure 15 for tidal effects help to clarify this. The point is that a combination of external forces on ocean and earth tide movements creates a trigger effect.

In these patterns, there is usually a noticeable twelve-hour and twenty four-hour rhythm of effects. In other words, the alternate rising and setting positions affect the disturbed area. With this, the changing position of the moon usually modifies the effect. The twenty four hour point would be a fairly close repeat in some cases but not if the moon was involved, as that would have moved about 13 degrees in that period of time. Nevertheless, with allowance for that some degree of foresight is possible.

In the 90-degree alignment that appears as an initial requirement, the position was as follows.

- Jupiter and Saturn were 107 degrees apart and finally at 90 degrees in March 1995. (The angle was closing).

- Venus and Jupiter were 75 degrees apart and exactly at 90 degrees on January 30 1994. (The angle was opening).

- Jupiter and Mercury were 84 degrees apart and exactly at 90 degrees on January 21.

- The angular position of Jupiter and Saturn is interesting, as it was prominent with sunspot cycles. It suggests that as the angle closed to 90 degrees that there would be more activity if reasonably strong grouping were also operating.

There were 12 earthquakes before 12:30 UT/GMT on the list for that date, in other parts of the world. Four were over magnitude 4. The noticeable factor is that they did not all match the rising or setting of the main group. Some may be foreshocks for the areas involved, or they may be aftershocks or an isolated shock.

These examples indicate that faults do react differently. Separate investigation could perhaps refine such details. The operative point is that with a diagram for any day any longitude and time can be plotted. With a variety of dates and reports, the patterns can be studied visually and computer models could evaluate probabilities. It is essential in such analysis to ascertain the near 90-degree positions of any planets to pinpoint the initial effect. Without it, the gravitational trigger effect would have much less impact. This would be because the core has not been stimulated and the magma has not become softer. The plates are then more rigid and the traction pattern has little or no effect.

The Second Day at Northridge

The second day, January 18 1994, covers a full twenty-four hours. To avoid confusion the numbers of earthquakes are presented in Table 12. The time frame can remain the same as the sun only "moves" one degree each day (it is of course the earth moving on its orbit around the sun). All the planets are virtually in the same position and only the moon would make any significant change. It moves on an average of approximately 13 degrees in one day. The rising and setting time is therefore roughly one hour later.

TABLE 12–THE SECOND DAY OF NORTHRIDGE EARTHQUAKES

Time GMT	Total for each hour	Time	Total for each hour GMT
00:00 – 01:00	13 (3 at 4+)	12:00 –13:00	2
01:00 – 02:00	5	13:00 – 14:00	5 (1 at 4+)
02:00 – 03:00	5	14:00 – 15:00	6
03:00 – 04:00	3	15:00 – 16:00	8 (1 at 4+)
04:00 – 05:00	4 (1 at 4+)	16:00 – 17:00	3
05:00 – 06:00	4	17:00 –18:00	0
06:00 – 07:00	3	18:00 - 19:00	3
07:00 – 08:00	4 (1 at 4+)	19:00 – 20:00	5
08:00 – 09:00	1	20:00 – 21:00	0
09:00 – 10:00	2	21:00 – 22:00	1
10:00 – 11:00	2	22:00 – 23:00	0
11:00 – 12:00	2 (1 at 4+)	23:00 – 24:00	1 (1 at 4+)
Total number of earthquakes listed	98		
Total number for Northridge	82		
Total for that area over Magnitude 4	9		

The most noticeable feature about the second day, is the jump in the number of shocks during the first hour to 13. This is when the group of planets, including the sun, were just setting. Any strong group that includes the sun tends to show a twelve-hour to twenty-four hour pattern. This then shows up as near the times of the rising or setting of the sun. If the strong group does not include the sun the times may vary but a similar pattern can still occur. Yet, apart from that the rest of the second day has intermittent disturbances. As the affected area moved round to the rising group of planets, there was an increase from the series of two shocks every hour. It jumped to 5, 6, and 8. Then as the area moved away from the rising position, it dropped again. There were one or two anomalies in the last few hours. The incidence of seismic activity for the last four hours (GMT) on the 17th, was 14, 2, 11 and 8. For the 18th, it was 0, 1, 0 and 1. This shows that there is not a strict repetitive pattern. Nevertheless, it supports the main premise that external traction can affect seismic faults.

The earthquake in Kobe, Japan struck exactly one year later and it was at 5:47 AM, just after sunrise. The diagram for that disturbance is of course not exactly like that for Northridge as the planets had moved. Nevertheless, the similarities and differences show that the broad principle cannot translate into a fixed rule. One extra effect for Kobe was that Venus and Jupiter were almost exactly ninety degrees from Saturn. As these three planets have the strongest radio noise, it implies that there would be an extra effect on the core of the earth. In addition, Jupiter and Saturn are essential planets in the sunspot effect. It was also at the time of Full Moon. In both cases, Jupiter had risen two or three hours earlier and as Jupiter appears to have a noticeable effect at the tangent gravity position, it is perhaps significant. For Northridge, all the external bodies were on same side of the earth as the sun. For Kobe, Mars and the moon were not. Other than that, they were very closely similar. The comparison helps to strengthen the indication that analysis could help to foresee earthquake cycles. Added to local observations a better degree of prediction might then be possible.

CHAPTER 7

BIOLOGICAL DISTURBANCES

The Secondary Effect on Animals and Humans

There are many cases of biological disturbances from natural electro-magnetic forces. These range from migraine headaches just before a thunderstorm to animal restlessness before earthquakes. In general terms, we say that it is the electricity in the air, or the electricity from the ground. In technical terms it is often much more complex. For example in physics textbooks electricity, magnetism, ionisation, microwaves and radiation may be under separate headings. Yet in natural phenomena, there may be no separation and all these effects are involved. Nevertheless, we can use general terms here as we are only concerned with the effects.

Animal disturbance before earthquakes is one such effect. There are many reports of such cases. One, very well documented, was before the big shock in Quetta in 1935. Quetta is in Pakistan, near the Afghan border, in a very mountainous region. The earthquake occurred on the last day of May, at three o'clock in the morning (21:32 GMT for the 30th). It was the beginning of the hot season, when people sit outside at the end of the day, and for this reason, the first effects were clearly noticeable. The domestic animals were restless and the birds flew away. Unusual electrical atmospheric phenomena occurred. Balls of light appeared. There was strange glowing at eleven o'clock on the previous evening, and for hours before the shock people were unable to sleep. The residents also said the animals were very restless about an hour before the impact.

There were no forewarning tremors. A diagram for that shock is of some analytic interest, showing a scattered set of alignments. Venus and Mars were 90 degrees apart some few days earlier. Two other pairs of planets made angles nearer to 80 degrees apart. The pattern did not seem very definite. Yet, it indicates how an astronomical system would have helped to foresee a critical time once the local signs were evident. For example at three o'clock in the morning, the moon was just rising and Jupiter was less than one hour off setting. Jupiter was relatively nearer as it was on the earth side of the sun. The shock occurred as the moon neared the horizon. Twenty-four hours earlier the angle between the moon and Jupiter would have been 155 degrees. The average angle with cross traction effects is approximately 145 degrees. The area was between the moon and Jupiter and these are prominent in gravitational effects. The horizontal compression from this would help to create electromagnetic effects. The earlier remarks about Yellowstone Park match this. Better knowledge could then have helped to indicate that this alignment could be dangerous, but at the time little was known about these associated effects.

Modern researchers would have no trouble explaining this, because they know that animal disturbance is by radiation from the ground. This is via ionisation from the rocks under pressure. Charged particles released by the ionisation also cause atmospheric electrical effects. Animal disturbance is a possible aid to early warnings of earthquakes, but unfortunately, it is not quite as simple as it sounds. As already mentioned there are reports of earthquakes with no prior animal disturbance and of animal disturbance with no earthquakes. What is more there can be similar effects away from a seismic fault area. This would be because the horizontal compression stretches across the adjacent land and is not just at the fault line. Nevertheless, disturbance is more prominent in seismic areas.

Researchers have also noticed that humans are also sometimes affected. This may show as headaches, general restlessness and irrational behaviour. Although most of these effects come from ground level radiation the whole atmosphere is affected, right up to the ionosphere. When we say that we live in the atmosphere of the earth, we should realise that it includes the magnetic field. Therefore, ionisation of the field can be by charged particles from the ground or from the ionosphere. Other forces can also affect the magnetic field and the planets at critical alignments would apparently play some role in this.

Electromagnetic Effects on the Nervous System

The effect of an increase in positive ions can create definite symptoms, referred to as the "serotonin syndrome". Serotonin is a neuro-hormone released from the pineal gland and it relates to nerve transmission processes, or in other words it helps in the function of the nerves. Distressing effects can apparently be offset by exposure to negative ions. Some reports say that serotonin is the key to rational thought and that when it is blocked there is a malfunction of the nerves that leads to difficulty and distress. Various drugs, including LSD, can affect the production of serotonin. Other types of research show that the nerves can carry both positive and negative ions. Sodium and potassium are apparently the most common but the balance can be upset so that the nerves carry all kind of ions. This occurs when there is exposure to different types of radiation.

The nerves are like a delicate electrical system. Overloading with an excess of energy, which it is not capable of handling causes distress. This leads to physical and behavioural effects. On this, it is of interest that almost all researchers are predominantly concerned with biological effects. Behavioural effects are more difficult to study. Research documentation mainly refers to biological effects on animals and the behavioural effects mainly referred to animals. From such reports, it might seem that humans are free from effects although they are biological organisms. However, it is much easier to observe animals under controlled conditions than to observe humans. Yet, if we take a lead from medical researchers, effects on animals can indicate similar effects on humans.

For example, research shows that electrical stimulation of the brains of monkeys could create behavioural variations. Ethologists discovered that the stimulation of nerve cells could affect the behaviour and the actions. Some monkeys showed aggression and assumed domination drives that were contrary to their normal character. A most significant point in these reports was that electrical stimulus above the normal tolerance level caused an increase in the type of expressed behaviour. What this means is that a neurotic person would become more neurotic, an aggressive or assertive person would become more aggressive or assertive and a violent person would become more violent, and so on. This is an extremely important point. If we apply the likely animal human correlation, it implies that there could be human, ie. social disturbance, at the times of increased radiation from the ground. Alternately there could be coinciding effects from magnetic changes. However, it should be noted that the researchers did also mention that contrary behaviour could manifest.

In considering the mechanism whereby such effects could operate, the technicalities of physiology, whether animal or human, are of interest. Sodium ions are part of the conductivity requirements. There are different types of sodium ions but the most common is sodium chloride as in common salt. This is extensively in seawater, which is a better conductor than fresh water. The significant point is that blood is almost as saline as seawater. We therefore have a system that is very vulnerable to electrical stimulation. The nerves can carry electrical energy in the form of charged particles and so can blood. Some people are restless just before it rains; this is particularly noticeable with children. Ions collect in the clouds and dissipate when it rains. The effect then diminishes.

There are also other causes. Hot dry winds can cause ionisation. Some reports of social disorder blamed an increase in positive ions. Israel installed large negative ion generators to combat the unrest. In parts of the Australian outback, the hot dry conditions sometimes cause depression and suicide. Perhaps it is a case for the use of negative ion generators. In Budapest, which is a mountainous and seismic area, the authorities fitted small negative-ion generators in police cars. Apparently, the increase of positive ions was blamed for the increase in social disorder. How the police felt when they left their cars was not mentioned!

One writer on Full Moon behavioural abnormalities specifically examined police records for correlation. Dr Lieber is a professional psychiatrist and obtained access to official police records in Miami, Florida. He presented his conclusions in a book entitled *The Lunar Effect*. His work clearly demonstrated that there is more violence at the times of Full Moon. In my own inquiries, I ascertained that the police attend more domestic disturbance calls at Full Moon than at any other time. In addition, mental patients are more disturbed at the Full Moon, and this is of course how we have the term "lunacy" or "Full Moon madness". The clue here is that basic characteristics increase at such times. The extent of the behavioural changes and the characteristics depends on the amount of stimulation.

The question that arises is why this should occur. Dr Lieber considered that it might be some type of biological tide effect. He considered that it might disturb the balance of hormone and chemical fluids in the body. Although his findings were not expressed in physics he did refer to ionisation, but his conclusions were in the direction of biology and psychology rather in terms of biophysics. The solution pointed to physics but this was not developed. Nevertheless, his research supported the principle of disturbance at the times of Full Moon. Before we consider the technical causes, it is relevant to add that hospital records show more admissions and more deaths at the times of Full Moon and apparently, more car accidents occur in this period.

The technicalities of the Full Moon positions are significant. The situation is that the magnetic field of the earth is not a simple sphere. It is pear-shaped, with a long "tail" that may extend as far as Jupiter's orbit. The solar wind causes this, that is to say the force of the radiation from the sun. This distorts the magnetosphere, depressing it on the side of the sun and creating the tail effect on the side of the earth away from the sun. At New Moon, the moon is on the same side as the sun, between the earth and the sun. This is the side where the magnetosphere is depressed. So in the New Moon position the moon is virtually outside the field of the earth and is not reflecting sunlight. At Full Moon, the moon is on the opposite side of the earth. It is then well inside the field but it is also reflecting sunlight into the field. As the earth and its field are generally negatively charged this means that the field is changed, becoming more positively charged because it is reflecting solar energy into the field. At such times, some people would react to this. The indications are that mentally disturbed patients are more disturbed. Violently disposed persons are more violent. Domestically tense partners become tenser. Reckless drivers become more reckless and sick persons with low resistance to infection become worse, hence the deaths. Ionisation from ocean-tide horizontal compression adds to this. Dr Lieber showed the incidence of high tide periods to local disturbances but did not define it in terms of horizontal compression.

Scientific Research on Electromagnetic Effects

Different literature revealed other effects. For example, the details collected by Sheila Ostrander and Lynn Schroeder in their book *Psychic Discoveries Behind the Iron Curtain* offer further insights into these problems. The title of this book is slightly misleading, as most of the discoveries were biophysical and psychic phenomena, as such, were not the main subject. As an example, Dr Andrews of Tallahasse, Florida showed statistical evidence to demonstrate that during surgery there was more difficulty with bleeding near the time of Full Moon. His report stated that 82% of excess bleeding was near these times. In this the trail again leads back to serotonin, for other sources of reference explained that serotonin is present within blood platelets in an inactive form. Blood platelets help to prevent the loss of blood by piling up at the source of bleeding. According to other sources, negative ion generators can help to reduce this excessive bleeding. This also implies that the bleeding relates to an increase in positive ions.

Other details of blood variations have also related to external causes. Dr Maki Tokata, of Toho University, Tokyo, found that the flocculation rate tended to improve about fifteen minutes before sunrise. Blocking of the rays of the sun by the moon tended to offset the effect. It also said that magnetic fields in a specific range affect the ability of the blood to coagulate. From such remarks, physics appears the best area for investigation. With some of these phenomena, the explanations were definitely biophysics. Dr Cavel Gulyaiev was able to demonstrate that the human being has a complex electrical field round the body. This is the "bio-field" or "energy body" but in older works, it is the "aura".

This was originally in the field of metaphysics which hard line scientists generally avoided. The gap is now closing and many explanations of strange phenomena are now in terms of biophysics. One experiment on the bio-field enclosed a person in an electrically insulated EEG chamber as a means of measuring the field. One woman had a field extending out as far as three and half meters. Checks by other researchers found that this was exceptional and with most persons, it is much closer to the body. Other researchers said that electrical stimulation causes the field to expand.

The significant point in such findings is that not only do we all exist in the field of the earth but we also have our own personal field. This implies that changes to the earth field will affect the personal human field. Soviet scientists say that there are three types of effect. One is personal, whereby the emotions cause changes in the field. Another is by the effect of mechanically produced fields and the third is from natural fields as from the sun and the planets. A point not made clear concerning the last type of effect was exactly how the effect operated. In a different experiment, telepathy was better when the sender and the receiver were in artificial fields. Ionised magnetic fields improved sensitivity in these areas. This matches reports of all types of persons being more affected at the time of Full Moon. As already explained this is because the earth field changes and becomes more ionised.

Other studies in more practical areas referred to plagues of locusts and rodents relating to sunspot activity and tree ring growth definitely matches the sunspot cycles. The operative point here is that as the cycle progresses there are more solar flares. These flares spurt out huge amounts of charged particles. As already described, they enter the earth at the poles, disturb the ionosphere so that there is a fall out from the ionosphere and also, to a lesser extent, penetrate the earth field and reach ground level. The measurement and analysis of these cascades is extensive. The effect in human terms is as dramatic as the Full Moon effect or perhaps more so as the positive ion input is temporarily greater. Statistical details showed that in the human range of effects there is an increase of suicides, psychoses, automobile crashes and deaths from heart failure. My own analysis of flares showed that they could occur any time and the effect would therefore be separate and extra to any Full Moon effects. Similar effects could therefore occur at any time and would relate to the specific cause.

Another writer whose work supports these views is Dr Lyall Watson. In his book *Supernature* he explains that a wide range of electromagnetic waves is continually bombarding man. He also draws attention to the effects from the moon. His work adds to the details described above and confirms that man does indeed have his own electromagnetic field. His writing endorses the point that the individual field changes by external influences. He quotes one authority (Burr) as showing that different species have a different electric potential. Kirlian photography has also shown that every individual has a slightly different field. It suggests that this difference is why one person reacts when another does not. The work of another investigator (Ravitz) has amazing implications. His research showed that the field of an organism resonates at the same frequency as the field in which it operates. This implies that changes in the supporting field would affect the personal field. The changes in the earth field, (caused by the external forces), could then affect the personal fields of all life forms living within it.

Dr Watson refers to the sensitivity of plants in these areas but the work of Peter Tompkins and Christopher Bird in their book *The Secret Life of Plants* provides more intriguing detail. They report that as far back as 1770 experiments in Turin by Professor Gardini showed that wires stretched near plants could affect them adversely. Recent experiments show that seeds in electrified containers grew more rapidly. This shows that there can be different effects, and in magnetic tests, there is a definite difference between a North Pole influence and a South Pole influence. Apparently, electrical stimulation or magnetic stimulation can be helpful or detrimental.

Professor Lund, in Texas, was able to measure the electric potential in plants and found that growth related to the electrical nervous system of the plants. This is of some interest here, as other observations had shown that both animals and humans are affected by means of their nervous system. One finding was that some plants were easier to monitor than others. Another report on the electrical responses of trees showed that there were changes in relation to the electrical state of the atmosphere. From this, it appears that electrical monitoring of plant responses might indicate subtle changes in ground radiation. This might perhaps have a practical application for earthquake warnings. It might be more viable than observing animals as there would be an objective instrument reading. Plant disturbances before earthquakes might then be a fact.

In the consideration of effects from ions, the work *Cycles of Heaven*, by G.I. Playfair and S. Hill, quoted the work of Aleksander Chizlovsky who demonstrated that sick animals responded to negative ions. Tests in Munich showed that positive ions adversely affected driving reactions whereas negative ions improved alertness. As well as this, there were more reports of deaths from driving accidents when there was a strong electrical atmosphere. In a collection of formal science papers edited by Professor Madeline Barnothy all these points were reinforced by conventional laboratory research. Most of the papers dealt with effects on cell and tissue but one or two mentioned behavioural effects. Overall, the reports indicated that there was distress. As with similar later research the effects were in two distinct phases. One was that the animals such as rats or rabbits, were distressed in the early stages, and the other was that there was nerve and tissue damage when the electromagnetic forces were beyond normal tolerance levels. It appeared to be a matter of degree, but some experiments showed a figure of 70% as a definite reaction. In addition, it explained that the effects of electromagnetic fields tended to persist after the main stimulus had ceased.

The most significant comment was a work by Professor Pressman, entitled *Electromagnetic Fields and Life* and the comment was that low intensity fields had more effect than high intensity application. Some reference was made to the possibility that planets may radiate such effects, but it was not emphasised because the work dealt mainly with laboratory research. The reports stated that the animals were distressed and agitated in the early stages. Other more recent works endorse all these findings. However, researchers were mostly concerned with animal responses and therefore observation of a change in psychological attitude did not come into the experiments.

Other recent research supports the findings of these different scientists but it is noticeable that in general the studies are concerned with effects on animals or the effects of artificial electromagnetic fields on human health. The possibility of effects on people, from natural fields, is not prominent in these reports, but there are references to the effect of low intensity fields. In their work, *Hidden Hazards,* Dr Laura and his colleague John Ashton stated in their discussion of microwaves that repeated exposure at low levels can have a delayed effect. We know that planetary fields are very weak in comparison to electrical appliances but Professor Pressman stated that such low fields do have an effect. Laura and Ashton's comments match this and show weak fields to have an effect. They mentioned that low levels of microwaves produced a range of nervous and vascular effects, and stated that, "studies have shown that the pineal gland is sensitive to electromagnetic field exposure". This is of particular interest as many scattered references indicate that the pineal gland is involved in certain types of nervous and behavioural instability, and relates to the serotonin syndrome.

In a different area, two other researchers, Oldfield and Coghill, consider the effects of the hidden mechanism of the brain. They state in their work, *The Dark Side of the Brain* that there is a need to study the biological field. From other clues, it indicates that this field is part of the mechanism whereby the nervous system is affected. The authors made two other significant comments. They said, "There is certainly an inter-penetrating electromagnetic field associated with all living organisms". This is important because of the indication that the greater field that surrounds it affects a lesser field. The other comment was that "very small energy changes in magnetic fields are of major biological importance". This is very interesting, as it is a further indication that living organisms can react to slight EMF changes. Of course, human beings are living organisms. Yet, scientists generally avoid this area; perhaps it is too controversial. Evidence is slowly mounting to demonstrate that there are genuine effects. The following example is unique.

The Charlotte King Effect

The researchers Rauscher and Van Bisc collaborated with Charlotte King in monitoring seismic activity. Charlotte King is extremely sensitive to electromagnetic variations and has the ability to sense developing seismic activity. Some of these effects are apparently very distressing and at the time of the Mt Saint Helens volcanic eruption, she experienced strong abdominal pains and suffered a minor stroke. The researchers presented their technical findings in a paper at the Tokyo International Workshop in 1993 under the title of "*Ambient Electromagnetic Fields as Possible Seismic and Volcanic Precursors*". These electromagnetic effects in the area surrounding a developing disturbance are the apparent cause of the biological distress. Measurements of these fields help in foreseeing seismic activity in the area. This shows that electromagnetic fields affect humans in many different ways.

Whale Beaching

One of the most interesting biological phenomena concerns why whales beach themselves for no apparent reason. Media reports usually give the times and places, and it is then quite easy to construct a seismic type diagram to see if the area was subject to horizontal compression at the time. In many cases, the patterns indicated that there could be an increase in positive ions so that it was a case of whale disturbance before earthquakes. However, observation showed that an earthquake did not necessarily occur in that area. Even so, there can be animal disturbance without an earthquake. The whales might then be in similar conditions. The whales might also be more vulnerable than land animals, because salt water is a good electrical conductor. As blood is almost as saline as seawater, the whales might be in an "electrical soup" from which they were trying to escape. This may not be the case but practical tests by measuring the ion count on the affected beach at that time would indicate whether there was an increase in positive ions.

CHAPTER 8

TWO UNIQUE EARTHQUAKES

Cause or Coincidence

The diagrams for the two major earthquakes in 1999, in Turkey and Taiwan, are different from the others and indicate extra potential causes. Astronomically the conditions for the two disturbances were very similar, but geographically they were quite different. In addition, the position in time was remarkable, as the second disaster was just over one month later. The comparative seismic graphs in Figure 21 show that the highest peak was almost at the same point in the lunar cycle. The explanation seemed to be one of ocean tide pressures in the monthly lunar orbit. However, difficult inconsistencies indicate other factors. In discussing them, it will be necessary to compartmentalise each point and relate them afterwards. These explanations are provisional and do not necessarily cover all the conditions that appear to contribute to a seismic disturbance. The details are therefore in two categories. One constitutes the astronomical facts and the other is my interpretation. Whether my conclusions are correct is a matter for ongoing investigation. In the end, the facts will speak for themselves but even so, understanding the principle and applying it in practical terms is a different matter.

93

An Estimation Formula

From the study of many diagrams, I developed a very general rule for anticipating seismic activity. The first step is to count the number of right angle alignments. In this a single 90 degree alignment graded as C, two were graded as B and three were graded as A and any further were A+. The closeness of external bodies as a group then indicated the possible trigger effect. In a similar alignment, the sun, the moon, Venus, Jupiter and sometimes Saturn were usually the most prominent. A group of any two of these would grade as 2, three would be 3, and so on. The most common strong combination was A3. This was for three near 90-degree alignments and a group of three of the prominent bodies. This is of course only a general "rule", because there are extra causes. Some of these are in the examples of planetary and tidal effects in Figure 22, but there are still others. Nevertheless, the general rule does give some indication, and further analysis will no doubt develop a better system.

As earlier explained, the hypothetical principle is that the combination of the electromagnetic radiation from any two planets at ninety degrees apart would stimulate the core of the earth. More such combinations would have a stronger effect. The final trigger effect is then from the traction effect of two or three external bodies in a close group. The combined gravitational force from these apparently stresses the fault as it moves into or out of the gravitational field of the external bodies. As just indicated, the strongest combination would be the moon, the sun, Venus, Jupiter and sometimes Saturn. However, this is somewhat rare and in general, three bodies combine to create a stronger traction. This is often when a New Moon falls in the same alignment as one of the planets mentioned. Even so, it seems that such a combination will have little or no effect if the core of the earth has not received stimulation from the other alignments, or some other cause. On the other hand, if more ninety-degree alignments apply to the core, the trigger may then be from only two external bodies or at times only one, usually the moon. The indication is that it depends on how much the magma has become fluid by the core activity. In addition to that, the proximity of the external bodies and possibly solar flares is important. We consider this in more detail later.

From all this is it possible to make some estimation of when seismic activity is likely, but it does not of course indicate where the earthquake will occur. We may be able to foresee a phase of earthquake activity, as with anticipating a cyclone season, but knowing where it will strike seems a more difficult problem. However, some foresight is certainly possible and by using the above principles, I was able to anticipate stronger activity in August of 1999 after the relatively quieter period in the previous months. Even so, the analysis of the Turkey diagram helps to show how difficult it can be to make an actual prediction.

Analysis of the Turkey Earthquake

To start with, Figure 23 for the 1999 Turkey earthquake shows there were four near ninety-degree alignments. These were Jupiter/Neptune, Uranus/Saturn, Uranus/Mars and Jupiter/Mercury. The only grouping was the sun and Venus, six degrees apart, or Jupiter and Saturn, thirteen degrees apart. The four alignments rated as A+and the sun and Venus counted as the 2. I graded the provisional diagram as A+2. In addition, the Moon could add its traction during its transit round the earth. However, if Venus were on the far side of the sun its traction would be less. Overall, the diagram is very scattered, and the "main rule" of most planets on the same side of the earth did not apply. Nevertheless, four sets of near ninety-degree alignments indicated the strong likelihood of extra activity.

The figures for seismic activity of magnitude 4.9 and above on the Richter scale during the previous months show a fluctuation of between one to four or five earthquakes each day. There are occasional peaks going to six or seven and sometimes peaks of ten or twelve earthquakes in one day. This raises the question as to why there were more earthquakes on those days. It is the higher peaks that often help to indicate why major shocks occur. This analysis therefore focuses on the peak for August 17 1999 and the Turkey earthquake, in the hopes that it may yield some constructive clues.

There were twelve quakes listed for that day. Six were in the affected area and six were elsewhere. In the Turkey diagram, the sun and Venus, which could be a potential combined trigger, are nowhere near rising. This puts the emphasis on the other general force of the moon and ocean-tide conditions. The positions of the sun and moon show that it was approaching a Neap Tide and in this case, Saturn and Jupiter were opposite the moon. This would add some slight extra force to the tidal traction of the moon, because the high tide is on both sides of the earth. Before considering the ramifications of this, it will be helpful to consider the differences with the other earthquakes on that day. In universal time, position and magnitude these were as follows: 2:58, 10 N, 85 W, 4.9, (2) 4:24, 28 N, 130 E, 4.9, (3) 4:31, 21 S, 174 W, 4.9, (4) 10:41, 29 N, 105 E, 5.1, (5) 15:06, 34 N, 32 E, 4.9, and (6) 19:56, 15 N, 119 E, 4.9.

The individual detail of these is below.

1. had the moon setting and Jupiter rising,

2. was indefinable,

3. had Venus and the sun setting,

4. had the moon rising with Jupiter and Saturn setting,

5. was also indecisive,

6. Venus and the sun were rising.

If we have diagrams for every seismic disturbance during any month, it will be clear that there are many variations. We are therefore only concerned with the principles involved. The central principle is that the combination of planetary forces affects the earth as a whole. In this process, many areas are disturbed. With the shocks on Izmit, in Turkey, the last 4.9 shock that morning was at 5:10 UT. This was well after the local sunrise, and there were some serious after shocks in the next few days. Geologically the seismic potential for that area is highly sensitive as the area is on a smaller plate sandwiched between larger plates. However, these are technical specialist considerations and our concern here is how the external forces have an effect.

The Geographical Conditions

The surface geographical conditions for that area are also unique and help to demonstrate how difficult it is to evaluate specific effects. Whereas with the sun, the surface conditions are uniformly similar, the conditions on the earth vary. Izmit is south west of Istanbul, on the eastern edge of a small sea called the Sea of Marmara. This is connected to the Black Sea by the Bosphorus, and to the Aegean Sea by the Dardanelles. The Black Sea has mountains to the south and to the east. The mountains to the south go down to the Mediterranean Sea and the Caspian Sea bounds the mountains in the east. These large expanses of water imply that the inland ocean tide forces would have helped to create the mountains by the regular horizontal compression from repeated tidal movements. In studying the position of the sun and Venus for earth or ocean tide effects, we can consider the Black Sea and the Caspian Sea. Izmit is 29 degrees east, and the eastern end of the Black Sea is 41 degrees east. The sun and moon would repeatedly create an ocean tide effect and although these shifting masses of water are critical, they are happening every day. It is the same with the consideration of external traction on the faults. As the earth rotates in and out of the gravitational fields, there are continual stresses. Scientific observation accepts that tidal forces do relate to some seismic activity but the cause of many differences is elusive. The scientists also say that the tectonic plates will drift in any case and local observation is the key to foreseeing a disturbance. Nevertheless, knowledge of the initial causes would obviously help.

In this present case, the area had recently escaped from the traction of the moon, which had set a few hours earlier. It therefore appears to be an example of an "elastic rebound", which geophysicists say does occur. My explanation is that this is an escape from external traction aided by shifting masses of water as the seas reassert their normal levels and the other opposing traction takes over. For example, in the diagram, the point of the disturbance is almost halfway between the traction limits of the sun and the moon. Jupiter and Saturn were well risen but there are no definite planets rising or setting on that area. Most diagrams tend to show some external bodies rising or setting at the time. For example, the earthquake in Turkey on November 12 1999 struck just after sunset. It was a typical example of contrary traction, with the sun and the moon on the setting side, pulling against Jupiter and Saturn on the rising side. The Turkey earthquake in August is different and indicates there are other possible factors. As the Taiwan earthquake graph also shows a similar peak in a similar period, a comparison will help to clarify the details.

The Taiwan Earthquake

The diagram for the Taiwan earthquake, Figure 24, had three ninety-degree alignments and the only relatively close group of external bodies was that of Jupiter and Saturn. It rated A2 on the estimation scale. However, the earthquake did not occur with these planets rising or setting, but almost overhead. The possible significance of this in terms of direct traction is part of this investigation and we can start by studying the generalities. There were ten earthquakes of magnitude 4.9 and above on that day, and only one was in a different area. This was at 9:32 UT, 46 N, 153 E, with a magnitude of 4.9 +. The area is that of the Kuril Islands, northeast of Japan, on the same plate as Taiwan. At that time,

Jupiter was almost rising and the moon was overhead. The rest, starting after the first shock at 17:47 UT with a magnitude of 7.7 were as follows: (1) 17:57, 6.0, (2) 18:03, 6.3, (3) 18:11, 6.1, (4) 18:16, 6.2, (5) 20:40, 5.2, (6) 20:43, 5.1, (7) 21:46, 6.5, (8) 21:54, 5.2.

For the main shock, the sun was not making any traction in that area, but Venus was about two hours from rising and the moon was just setting. This seems to be the real trigger and again focuses attention on the effects of tidal movements. The other aftershocks appear to be the result of different bodies affecting the area, and the changing force of the ocean tides. They help to show how once there is a disturbance any extra force can create an effect. This indicates how prediction, either before or after the main shock, is by no means reliable. The twelve and twenty-four hour patterns are easier to follow, as with the Northridge sequence of earthquakes, but scattered alignments are not so readily analysed.

The Geography for Taiwan

Taiwan is part of a chain of islands that continue south from Japan to the Philippine Islands and then to New Guinea. This is in a long seismic area with the whole weight of the Pacific Ocean adding its thrust to the eastern coast of the main landmass with every rotation of the earth. There is a range of mountains down the centre of Taiwan, like a long spine. This indicates mountain building from horizontal compression caused by the tides and the drift of tectonic plates. The mainland has no mountains near the coast and the highest part is near Kimming. These mountains appear to relate to the tidal movements in the South China Sea and the Bay of Bengal on the western side and to some extent from water in the Gulf of Siam. These constantly shifting tidal forces exert a regular pressure over a long period. In some areas, they create bores where the water rushes in like a tidal wave, and in other areas, they create rip tides and swirling currents. Although the moon is the main cause of this water movement, the nature of the land determines how they affect the landmass. The position of the fault line particularly determines where pressures may be most effective.

In this situation, we have an island with a large landmass to the west and the Pacific Ocean to the east. The ocean tides are obviously the source of a strong horizontal pressure. When the moon was overhead, it would have helped to raise the high tide mound of water in the Pacific. As the ocean area rotated away, the water would try to stay beneath the moon and would press against the eastern coast. Taiwan, as an island, would have extra water on its island shelf, but again, this type of effect is happening regularly. Nevertheless, with other contributory factors this pressure becomes more effective. We therefore have to look for these extra contributory causes. In the tide raising process, Jupiter and Saturn would be adding some direct traction and exact measurement of tidal levels at that time could help to clarify any differences. As pointed out earlier, these diagrams have no allowance for the inclination of the earth and the exact time of rising or setting has to be calculated, or obtained from tables. From corrections, it appears that the moon was much closer to setting at that time. This makes a standard pattern although the moon alone does not usually appear to act as a trigger unless other factors have contributed. However, we have to remember that the external traction is not just on that area, or only the ocean, it is on the plate overall. The moving tide mound may cause other movement on the plate, such as for example a slight rocking action as the tide mound shifts. Furthermore, geophysicists say that there are even movements in the magma caused by lunar and solar tidal forces. From this, we can see that a consideration of local surface conditions may be far too restrictive and we have to widen our investigation.

The Heliocentric Positions of the Planets

In a search for contributory causes, there is the question of the exact position of the planets at the time of the two earthquakes. To answer this we can examine the heliocentric position of the planets. The diagrams in Figure 25 and 26 show the positions. The diagrams show the earth, number 3, below the circle, to give a better impression of the view from the earth. Both diagrams show that all the planets are on the same side of the sun and all the traction is pulling in the same general direction. This indicates that there is a broad tidal effect. More than one writer has asserted that there are tidal effects on the sun, but this has not gained much support because the explanations were not specific. As with seismic disturbances, the focus was always on the direct traction of a high tide position. Here we are considering the question of whether indirect traction, at the indirect traction position, could disturb a sunspot and generate extra energy that affected the earth. From the diagram, we can see a definite area that is subject to contrary traction from one group of planets setting and another group of planets rising on the area facing the earth. On the first diagram, the area is within an arc of about seventy-five degrees. In the other, the area is nearer one hundred and twenty degrees. The question is, could these alignments have caused a flare that affected the earth, or could there be an extra effect. We know that flares affect the ionosphere and help to cause the polar lights but extended effects from this seem difficult to prove. Even so, it is necessary to examine them, in case there is an extra clue.

Our immediate interest is whether there were critical alignments before the times of the two earthquakes. That is to say, whether alignments could disturb a sunspot and cause a flare, that in turn contributed to seismic activity. Although there is some indication of a connection, as the later flare graphs show, there are often flares with no seismic activity, and seismic activity without flares. In addition, there are other contributory causes. The problem is doubly difficult as we are attempting to synchronise two different processes. Each process is part of the puzzle, and we must therefore look for any extra matching phenomena.

As an example of how difficult it is to demonstrate a connection, we can start with the sunspot records and the two graphs in Figure 27. These are typical of monthly graphs. Analysis shows that sunspot graphs and seismic graphs make a broad match in a smoothed curve over a long period. The graphs in Figure 27 show a peak in sunspots at the beginning of the month, well before the Turkey earthquake but the peak was only five days earlier for the Taiwan earthquake. With such differences, it is not clear enough to point the finger at the sun and say, "there is the culprit". The long-term graphs show that there is a similarity, and suggest there is probably a mutual cause. The signs point to the planets, which appear to affect the core of the sun and the earth in different positions. Nevertheless, the investigation indicates that suitable alignments do not occur at the same time for the earth and the sun. Sunspots, earthquakes and flares therefore show different graphs that can never have an exact match. Although the principle is the same, the application is different. Before we consider that, there are other causes of gravitational stress, such as proximity, which are worth examining.

The Relative Position of the Planets

To consider any extra gravitational effect we can examine the actual position of the earth itself in relation to the main effective bodies. Figures 28 and 29 show the relative positions of the effective bodies at the time of the earthquakes. They give the relative positions of the sun, the moon, Venus, Jupiter and Saturn and the table below shows their distance from the earth at the different times.

TABLE 13–DISTANCE FROM THE EARTH ON THE 17^(th) OF EACH MONTH 1999
(Millions of kilometres)

Month	Sun	Venus	Jupiter	Saturn
June	152. 0	100. 0	811. 2	1488. 4
July	152.0	65.3	747.6	1427.8
August	151.5	43.3 *	677.8	1352.7
September	150.4	56.3	620.1	1281.9
October	149.1	87.0	593.5 *	1237.1
November	147.9	121.8	607.4	1230.1 *
December	147.2 *	154.1	656.7	1264.1

Earth distance from the sun in 2000: January 147.2, February 147.8, March 148.9

The asterisk indicates the nearest positions.

The table shows that Venus was at its nearest in August, Jupiter was at its nearest in October and Saturn at its nearest in November. The distance of the earth from Saturn then increases rapidly. The earth is almost nearest to the sun in December, and is at Perihelion in early January. In this position, the earth is moving at a greater velocity, and conversely at its furthest point of Aphelion (in the beginning of July) is moving more slowly on its orbit. Suitable alignments from the slower planets would also last longer. The significance of this increased or decreased orbital velocity might perhaps show up in computer analysis or extended graphs. The fact is, the earth was at the slower part of its orbit in August.

In the first of the two diagrams, for Turkey, Venus is very close and the earth is at a position in which it appears about to fly off at a tangent towards Jupiter and Saturn. We have Venus almost in line with the sun and the traction from the sun and Venus would combine. At the tangent position, the earth is moving towards the combined mass of the two major planets, and we can assume there would be a Doppler effect. This would be in two areas; one is that the electromagnetic radiation from the planets would increase. The other is that the gravitational force would be stronger. This may be debatable, as gravity is not measurable in the same way, however the diagrams do indicate an unusual configuration capable of mathematical evaluation in conventional physics.

The second diagram, for Taiwan, is equally interesting. In these movements, we have to remember that the earth is rotating and the moon is affecting the ocean tides. The major planets may also affect Venus and the moon at their nearest, but whether this is significant for the earth is difficult to ascertain. Another interesting point is that both Jupiter and Saturn became retrograde at the end of August, approximately half way between the times of the two earthquakes. This is an apparent motion whereby the planets appear to go backwards against the star background and is the effect from movement of the earth on its own orbit. In terms of relative motion, the two planets would temporarily appear to stand still. Whether this has any importance in terms of gravitational force is a matter for specialists. Conventional descriptions generally refer to distance and movement, but this is a point of change from one movement to another, or more specifically, no apparent movement. This question of the relative motion also applies to Venus at that time. In the official almanacs, Venus is retrograde during August. During this period, until Venus was in line with the sun on August 20, the distance between Venus and the earth would be closing. However, Venus would not be in the same relationship to the earth at the time of both earthquakes. The daily distances of Venus in August, in millions of kilometres, were as follows: 17th 43.3, 18th 43.2, 19th 43.1, 20th 43.1, 21st 43.1, 22nd 43.1, 23rd 43.2, 24th 43.3.

The change of apparent motion and gravitational effect from Jupiter and Saturn may be more important at the time of the Taiwan earthquake. With the rotation of the earth, we could also have a critical position for one area, ie. one tectonic plate. This would be analogous to that of the earth on its orbit. In other words, the seismic area is swinging towards an external force and then pulls away from it. The lines of force would determine where the maximum strain would be. One interesting similarity in the two diagrams is the position of the moon. In both cases, it is "behind the earth". That is, it is on the side away from the direction in which the earth is moving. Whether this could have effect is not clear. However, these problems occur in engineering and tests to find the breaking strain or yield point are standard routine, so a practical model seems quite feasible.

The Moon

The moon did not appear to be at its nearest position to earth, as the following figures show. These are the distances of the moon from the earth, in thousands of kilometres, at seven-day intervals:

TABLE 14–DISTANCE OF MOON FROM EARTH FOR AUGUST & SEPTEMBER 1999

(Thousands of kilometres)

August		September	
1st	378.8	5th	371.6
8th	366.7	12th	394.0
15th	391.9	19th	399.5
22nd	400.2	26th	368.3

On 17 August, the distance was 400.6. On 20 September, the distance was 395.6.

Neither of these positions appears significantly nearer but the orbit of the moon is at an angle and varies in celestial latitude as much as five degrees north or south. In addition, the earth is inclined at an angle to the ecliptic of approximately 23.5 degrees. Tables for the position of the moon allow for this by showing the declination. This is the angle of the moon, north or south and varies from zero to twenty- eight degrees. The point here is that irrespective of distance the moon could sometimes affect the ocean tides differently. As an example, an eclipse occurs when the moon is at zero degrees celestial latitude and is on the plane of the ecliptic. The gravitational force of the moon is then directly in alignment with that of the sun. This also occurs with planets, when the moon occults the planet. The difference may not be significant with planets but the exact alignment of the sun and the moon stands out in comparative analysis. However, the moon is not necessarily at Perigee at such times.

The upper and lower positions of the lunar orbit, at the extreme northern or southern point, might therefore temporarily be more effective in a particular tide-raising situation. The variations of this, in relation to the inclination of the earth, are multiple. This variation, of the position of the earth, in relation to the external bodies, may be one of the reasons why definite traction alignments affect one fault and not another. In other words, the lines of force are not suitably in line. A comprehensive computer analysis may be able to resolve some of these problems. Hypothetically, it may be possible to calculate times when San Francisco, or any other area, is in the same angular position with similar alignments to those at the time of an earlier earthquake. No doubt, a mathematician could calculate near repetitive conditions to some extent. This would then indicate whether the area was subject to the same lines of force as well as equal traction. Like all the other points that demand formidable expertise, it is indeed a task for specialists.

Convection Currents

There is one further point that appears to relate to planetary alignments. This is the development of convection currents. Both solar physicists and geophysicists refer to convection currents. As explained earlier, the currents in the sun rise to the surface and form the sunspots. In the earth, they occur in the fluid magma, but of course do not reach the surface in the way they do on the sun. Even so there may be a connection. For if, the effects from Jupiter and Saturn cause the convection currents in the sun, how do they originate in the earth? The earlier conclusions suggested that the changing polarity of the sunspots related to the changing positions of Jupiter and Saturn. This implies that the energy of these planets could cause the convection currents of the sunspots. If this is so, we can assume that the planets could equally cause convection currents in the earth.

Although there was no definite indication of specific planets relating to the earth, as with the sun, the principle appears to be the same. The earlier investigations did indicate that Venus and Mars appeared to make similar alignments, but the details were not firm enough. Nevertheless, in Chapter 6, there was the report that unusual low frequency electrical surges occurred two weeks before the Northridge earthquake. It was at the time when Venus and Mars were exactly in line, and suggests a possible similar process as with the sun. If true, the observation of geophysical phenomena with alignments of Venus and Mars might indicate what happens, as in the sun with Jupiter and Saturn. However, it may be that any two planets in a suitable alignment create convection currents in the mantle of the earth. These currents would probably cause the magma to become softer and an earthquake would be more likely. Presumably, this would show in local tests for electromagnetic variations. Much of this probably overlaps on current knowledge but the consideration of the planets as the initial cause could perhaps clarify the situation and help with extra warnings of likely activity.

A simple indication of some connection is the comparison of ninety-degree alignments to the sun with sunspots relative numbers. This was in Table 6, pointing out that suitable alignments with any two planets did match. The indication here is that if more ninety-degree alignments mean more convection currents and sunspots, then more convection currents in the earth could mean more earthquakes. Hypothetically, the formation of currents in the mantle would be similar to that in the sun. This means that if Jupiter and Saturn create four sunspots, two in the north and two in the south the extra alignments would create extra sunspots in that framework. However the sunspots are consistently positive and negative, and this raises the suggestion that Jupiter and Saturn may be the dominant factors, or that any two planets of opposite polarity can create an effect. Generally, more alignments appear to match an increase in sunspot relative numbers. If the same applies to the earth, more alignments to the earth would be critical.

One other interesting phenomenon may relate to the formation of convection currents in the earth. Cosmic rays and other high-energy particles approaching the earth do not just penetrate the atmosphere but follow an ordered path. The descriptions say the positive ions spiral round the earth in one direction and head towards the poles. In contrast, the negatively charged particles spiral round the earth in the opposite direction and go towards the poles. This implies that there are two energy streams, spiralling in a different direction. In this way, they could presumably create a pair of convection currents in the same way that they develop in the sun. If we apply this principle to the sun, we have to ask where that energy influx originated. The hypothesis was that the positive and negatively charged planets emanated energy waves that directly affect the core. In a different view, it may be that energy enters the poles of the sun. In such a process, a solar flare directed towards the earth could perhaps act as an extra boost to create an earthquake. In this case, the other planetary alignments as shown in the heliocentric diagrams would be the means of disturbing a sunspot and the questions about flare energy would still be valid.

A Summary

In summary, the indication is that the planets have a similar effect on the sun and on the earth during the same period. That period is the sunspot cycle. What appears to happen is that slower moving planets make a suitable alignment for a longer period and therefore dominate the long-term effects. For example, when suitably aligned, Uranus and Neptune create the continual production of convection currents over a long period. Within this steady stimulation, Jupiter and Saturn add their effect over a shorter rhythm. Superimposed on that are short-term effects, caused by the other planets. This seems to be the basic format. The tidal effects that act as a trigger for flares or earthquakes are extra, and furthermore seem to have less effect if the first cause is absent.

We can now consider peaks in seismic activity, with peaks in sunspot activity. As already stated, the initial alignments affect the earth and the sun independently. The arrangement of the alignments may therefore cause a peak of activity in the sun and a peak in seismic activity but they do not synchronise. At one time, the sun may be stimulated first and at another time, the earth may be stimulated first. The relationship of peaks in the comparative graphs would then be different. If peaks in solar activity were always first, we could assume the sun caused seismic activity, but this is not the case, as planetary alignments appear to affect both the earth and the sun independently. Furthermore, they also affect each other. In this system, the earth can also affect the sun. It is not just a case of the sun and the earth but of the whole solar system in a changing play of forces and energies, that keeps the whole mechanism in motion.

The basic pattern of energy exchanges via the ninety-degree alignments seems simple enough but the gravitation modifications appear very complex. Everything seems to depend on these two principles, which have potential for examination under laboratory conditions. In the end, there should be nothing speculative about it. To narrow down the field of inquiry we will consider some of these points further on.

CHAPTER 9

THE 2000 SUNSPOT MAXIMUM

Sunspot Predictions

An IPS Radio and Space Services Solar report obtained via the Internet in July 1998 included an updated status report entitled *Prospects for the Future Solar Cycle 23*. The report stated that the Sunspot Minimum of the previous cycle was May 1996. The details then followed that the current cycle, No 23, was likely to be " a larger average cycle". This means that higher than usual Sunspot Relative Numbers are likely. The estimated date for the next Maximum was March 2000. However, the report stated,

> "The art of predicting the magnitude of sunspot cycles is still rooted primarily in empirical relationships… it must be noted, however that predicting the month of the maximum is even less certain than the sunspot number. The actual month of sunspot maximum could be as early as January 1999, or as late as June 2001."

The simplest method of estimating the likely date of the next sunspot maximum is by deducting the length of the fall to the recent minimum from the average length of the cycles. An analysis of previous patterns can refine this but still leaves the prediction flexible because there is no known timing factor. Hypothetically, any cyclical system would have some degree of repetition and similarities in its process. We can therefore examine the position of the planets for the year 2000. On this basis, the date of May 11 is a good point of reference. This is in Figure 30. In terms of the previously explained principles, this is a useful period for considering both solar and terrestrial effects. The connection between these two activities arises because at that time, the major alignments almost match for the earth and for the sun, and therefore it is a useful hypothetical date for the analysis. However, that date has another interesting critical point, in terms of the diagram in Figure 2a, because the relative heliocentric positions of Jupiter and Saturn are unusual. As a "left hand" or "right hand" pattern the sunspot maximum is usually *after* these two major planets have made the ninety-degree alignment and *before* the conjunction or opposition. When the pattern was irregular and "missed a beat" it was after the conjunction or opposition positions, and therefore any irregularity would perhaps involve two cycles as part of the overall "22 year cycle". This would appear to start when Jupiter and Saturn were in the same alignment.

The suggestion from this visual interpretation is that the ninety-degree alignment is a critical factor, as the process appears to generate maximum energy at that time. There is then a time lapse as the effect reaches its peak; this is before the conjunction or opposition because that is when the half cycle is complete ie. from conjunction to opposition, or opposition to conjunction. The next process is the development of new sunspots. However, there is always a delay in these processes as the new convection currents take time to produce the new sunspots. At this stage, this is hypothetical but the advantage of studying astronomical processes is that the facts eventually clarify the situation.

The exact heliocentric positions of the planets in March 2000 are also very interesting. The tables show that on the 21st Jupiter, Mars and Saturn are within a five-degree arc and Neptune, Venus and Uranus are within a thirteen-degree arc. These two groups are almost ninety degrees apart. As this seems unique, the heliocentric positions to the nearest degree are worth examining in detail. In the first group, we have Jupiter 43, Mars 46 and Saturn 48. In the second group, we have Neptune 304, Venus 309 and Uranus 317. (The earth is 180 and Mercury is 227). The two groups create a total of nine very near ninety-degree angles. Hypothetically, this would match a good production of sunspots. They would presumably increase after the two faster planets, Venus and Mars have moved into a close alignment. In addition, when there are more sunspots there are usually more flares and a note on solar storms may here be relevant.

Solar Storms

Scientists are interested in solar and geomagnetic storms because they affect many technologies, from high voltage power grids to communications and radar, and consequently there are continual reports. One report, in March of 1999, stated that a solar storm was developing on the sun and there was a comment to the effect that it was at almost at the same time as the previous year. Comparative diagrams showed the 1999 diagram was very similar to the 1998 diagram, with the earth in a similar relationship to the major planets. The possible significance of this is because the earth is also an effective planet in this scheme. The faster planets were then again in a position to add to this and make cross traction with the slower planets and with each other. Hypothetically, the position of the solar storm might relate to planets rising or setting on that area or it may be the product of ninety-degree alignments stimulating the core. In the period for March 2000, the planetary positions have an extra interest in relation to the official prediction. In diagrams for this period, there are two areas where cross traction areas on the earth side might cause flares that would be visible from the earth. In the diagrams, the faster moving planets, ie. Mercury and Venus are different but still effective and the earth, as an effective planet, is in the same place. Comparisons with the earlier chapter on flares will help to clarify this and we consider he problem again with the graphs. The question that arises here is whether planetary traction can do more than disturb a sunspot, ie. it may help to create a solar storm. It is unwise to blame everything on the planets but we have to include them as suspects even if only to eliminate them from our inquiry.

The Sunspot Maximum

The figures given above show that in March Jupiter and Saturn are five degrees apart and the exact conjunction took place at the end of June. The Maximum is usually before the conjunction. June would hypothetically be the last period for a regular Sunspot Maximum. As it was after that, it could be an irregular cycle. The indication appears to be that the timing of the change over of the polarity of the sun and the sunspots relate to these alignments. Jupiter and Saturn were in direct alignment at the end of June of 2000. Further details are in Figure 31. According to descriptions, the new sunspots appear while some of the old ones are still in existence. This means that new spots of a different polarity will develop in high latitudes while the others, in the lower latitudes are trailing off. Observation will eventually clarify the process, because we can compare the appearance of the new spots after the heliocentric conjunction of Jupiter and Saturn. However, confirmation of the detail is often later. Even so, the whole process is so mathematically calculable that all the conclusions are open to hard investigation. In view of the large number of ninety-degree alignments in March we have two interesting periods, but the one in May appears to be the final set of alignments in this cycle.

Associated Seismic and Solar Phenomena

The interesting question is, are there any observable differences in such a cycle? In considering the possible significance of stronger than usual solar activity, the overall requirements for solar and seismic activity would have to match against the planetary possibilities. Figure 22 shows the general pattern for effective seismic disturbance to occur. We have considered the detail earlier and the diagram is self-explanatory. The general requirements for seismic activity are as follows:

- A near ninety-degree alignment of two or more planets activates the core.

- A group of external bodies making a combined gravitational field stresses the fault as it rotates into and out of the field.

- All, or most, of the effective external bodies will be on the same side of the earth.

- An extra input of energy into the earth via solar storms or flares activate the core.

- Extra horizontal compression from slightly higher ocean tides stresses the fault.

The above principles are only partly applicable to the sun. There is a planetary tidal effect but not exactly as with ocean tides. The principle of the core stimulation is apparently the same:

- Two or more planets at a ninety-degree alignment affect the core of the sun.

- Jupiter and Saturn are most prominent in this process, as they also appear to relate to the polarity changes of the sun.

- Two or more planets making cross traction may disturb a sunspot and cause a flare.

The year 2000 is unique, and expert analysis of effects would no doubt yield many clues. In the search for previous similar alignments, December 21 1699 stands out. According to historical records, some significant earthquakes in that period included Cateria, Italy, in 1693 (60 000 killed), Hokaido, Japan, 1730, (137 000 killed), Calcutta, India, 1737, (300 000 killed), and Lisbon, Portugal, 1755 (70 000 killed).

In studying any period, the major slow moving planets are in the critical alignment for a longer period. In addition, the apparent retrograde motion, caused by the movement of the earth, sometimes creates a repetition of the alignments as viewed from the earth. As well as this, the faster moving planets can make suitable alignments even if only for a short time. The result is that there is not one short period of strong activity. Instead there is a slow build up as the major planets close to a suitable alignment, and within this slower period are fluctuations caused by the faster moving planets making short term alignments. Woven into this pattern is the grouping and dispersing of strong traction bodies that act as the trigger by jolting the fault as it encounters the limits of the traction field. In addition, there is the proximity of the planets and other external factors, such as perhaps the effect of solar flares.

Sequential diagrams for any period, for example 1700 to 1760, show variations. With modern methods, we can analyse these against solar and seismic changes in activity. The broad principle shows in the following list of planetary positions. The Right Ascension of the planets for noon of December 21 1699 is as follows:

TABLE 15–RIGHT ASCENSION OF THE PLANETS, IN HOURS
AND MINUTES FOR NOON OF DECEMBER 21 1699

Sun, Moon & Major Planets	RA hours/minutes	Major Planets	RA hours/minutes
Sun	17:59.5	Jupiter	18:34.6
Moon	18:02.0	Saturn	22:02.1
Mercury	19:12.6	Uranus	06:40.1
Venus	18:40.2	Neptune	00:12.4
Mars	13:08.5	Pluto	09:23.1

Many disturbances occurred in early 1700. The alignments were unique in 1699, with six near ninety-degree angles. They were Neptune/Jupiter, Neptune/Venus, Jupiter/Mars, Mars/Uranus, Neptune/Uranus, and Mars/Mercury. They vary from 82 to 85 degrees apart. According to the hypothesis, the indication is that the core would be very stimulated at that time. Hypothetically the softer the magma is the easier the plates would slide. The external traction could then affect the fault more readily. The close group of celestial bodies that could act as the trigger is the sun, the moon, Venus and Jupiter. These are the four strongest traction forces. They were all within an arc of ten degrees, and it was the day of the New Moon, which makes it comparative to May 4, 2000.

This 2000 period has a different grouping and there are unusual complexities. In the example above, the period of the Maunder minimum, discussed earlier, was from 1645 to 1715. The list of alignments in Table 15 is therefore right in the middle of the period. The nearest sunspot maximum to the years listed above, are 1738.7 and 1750.3. Neither of these times had the close alignment of May 2000, however, there were some critical alignments a few months before the spate of earthquakes. These indicated that the major slow moving planets do match with protracted periods of seismic disturbance. Yet, overall the most interesting set of alignments was for 1699. In view of the Maunder minimum, any attempt at drawing a conclusion from comparative analysis seems difficult. If, in that case, the decrease in sunspot activity coincided with an increase in seismic activity, we can draw some conclusions. One is that any irregularities in the sunspot cycles might have a different extended effect on the earth. Alternatively, we could conclude that independent planetary effects on the earth are the significant factor. In that case, a connection with sunspot activity is not necessarily a strict requirement for seismic activity.

What emerges from these attempts at comparison is that the 2000 period stands out as one of unique interest. As modern methods include meticulous observation and a wide range of recording, the analysis of phenomena in the year 2000 will undoubtedly be of special interest. Despite the general similarities, there are other practical points that have to be included in a comparison, such as the proximity of the effective bodies. Planetary tables for May 2000 show Jupiter and Saturn, as well as Venus and Mars on the far side of the sun, and they are more or less at their maximum distance from the earth. The positions in December 1699 show that Jupiter and Saturn were very distant at that time although the earth was almost at its nearest to the sun. In contrast, the earth was on the same side of the sun as Jupiter and Saturn at the time of the Turkey and Taiwan earthquakes. In 1999, they were then relatively near, and therefore presumably had more effect. With such conflicting details, we have to exercise caution in trying to draw conclusions. Because the planets and moon are all on elliptical orbits, their distance from the sun and each other constantly varies. There may be times when the earth could be in a critical period but the point here is that May 2000 is not an exact replica of earlier positions and an assumption of duplicate effects is unwise. Even so, the alignments are unique in terms of possible irregularities. The Right Ascension positions and heliocentric positions of the planets for May 11 2000 are below (Figure 30 shows the geocentric positions).

TABLE 16–RIGHT ASCENSION AND HELIOCENTRIC POSITIONS OF THE PLANETS FOR MAY 11 2000

Sun, Moon & Major Planets	Geocentric RA - hrs/mins	Heliocentric – in degrees	Major Planets	Geocentric RA - hrs/mins	Heliocentric – in degrees
Sun	03:12.7	-	Jupiter	03:05.4	48
Moon	10:13.2	-	Saturn	03:12.7	20
Mercury	03:21.4	60	Uranus	21:33.4	317
Venus	02:35.5	30	Neptune	20:35.6	304
Mars	04:12.1	74	Pluto	16:48.4	251

When considering these details, it is important to remember that we are attempting to relate two sets of alignments. These are the heliocentric positions for the sun and the geocentric positions for the earth. Against that, we are trying to match two sets of events, ie. possible irregularities in the sunspot cycle and extra terrestrial activity for the earth. It is therefore likely that there will be differences even if there are similarities.

Earthquakes and the Recent Sunspot Minimum

The sunspot minimum for May 1996 had a low smoothed relative number of 8 and it is interesting to compare this with other periods of sunspot minimum. The smoothed numbers at minimum vary from 0.0 to 12.3 with an average of 6.0. The largest at sunspot maximum vary from as low as 48 to 201.3, with an average of 112.9. The 201.3 maximum was in 1957.9 and it rose from a low 3.4 in 1954.3. From this, we can see that attempting to predict the degree of activity from numbers alone is very difficult. When we try to translate this into likely seismic activity the situation is even more complex. The two graphs for August and September 1999 make it quite clear that the sunspot numbers and the earthquake numbers do not rise and fall at an equal rate. We have already considered this, but for an extra consideration of the current phase, we can make a further comparison. We can match details of seismic activity for May 1996 with that period of sunspot minimum activity, using the figures for earthquakes of magnitude 4.9 and above.

Although these comparisons indicate a general rise and fall, there are always discrepancies. As indicated earlier, the cause is the difference in time of the planetary alignments. Even with only the two essential planets of Jupiter and Saturn, we can see quite easily that they do not make an effective alignment at the same time for each focus. That is to say, with the sun at the focus the ninety-degree alignment will be at a different time from the earth at the focus. These differences appear to be part of the reason why the peaks do not match up in comparative graphs. With the faster moving planets also adding an effect, together with the sun and the moon acting as a tidal trigger for earthquakes we are unlikely to have an exact match of seismic activity with solar activity. The intrinsic point seems to be that the sunspot activity alone does not create seismic activity, but develops parallel with it. There does appear to be some contributory cause from the solar energy that affects the earth and these figures match the general patterns.

TABLE 17–SUNSPOT RELATIVE NUMBERS & EARTHQUAKES
FOR SUNSPOT MINIMUM–MAY 1996
(Magnitude 4.9 and above for each day of May 1996)

Day	Quakes	Rel Nos	Day	Quakes	Rel Nos	Day	Quakes	Rel Nos
1	4	0	12	0	16	23	2	0
2	6	0	13	4	15	24	1	0
3	9	0	14	4	14	25	0	0
4	4	0	15	2	11	26	1	0
5	0	0	16`	0	9	27	2	0
6	2	11	17	0	8	28	1	0
7	7	12	18	2	8	29	1	0
8	0	13	19	3	0	30	1	0
9	2	12	20	0	8	31	2	0
10	2	14	21	0	0	Totals	69	169
11	7	18	22	0	0	Mean	2.2	5.5

As can be seen the sunspot relative numbers rose rapidly from the 6th to the 11th of May 1996, and then fell steadily to the 19th. In contrast, the earthquake figures jumped up and down in the first part of the month and dropped, with variations, to two or less after the 20th. The last ten days were therefore a very close match of little or no sunspot or seismic activity. Random monthly comparative graphs only show a general similarity. As already explained, there will never be an exact match on sunspots and earthquakes alone, because of the other variables. The serious question here is to how far sunspot activity affects seismic activity and how far they develop in parallel because of the same basic cause, ie. via the planets.

TABLE 18–POSITION OF MAJOR PLANETS FOR MAY 15 1996

Major Planets	Right Ascension	Heliocentric Coordinates
Jupiter	19:08	278
Saturn	00:16	359
Uranus	20:16	301
Neptune	19:48	296

The positions of the major planets for the sunspot minimum in 1996 show that Jupiter and Saturn were still 81 degrees apart in the heliocentric coordinates. However, the cycle had passed the ninety-degree position and the maximum sunspot period. This alignment indicates that there would still be some time before the conjunction of the two planets. At other times of sunspot minimum, the angular distance between these two planets is different. That is to say, the completion of the presumed change of the polarity systems of the sun could take place nearer the actual minimum. This may show in comparisons of the timing of the change. In other words, the angular distance determines how much more time it will take to complete the change.

If we study the positions for each minimum in Figure 2b, we see much more clearly that the peak energy relates to a regular set of alignments at maximum. There is then a delay in the effect, followed by a decline in activity as the planets move towards a conjunction or an opposition. After completion, the whole process starts again. The differences in timing appear to be from the variations of the elliptical orbits and other planets that are part of the system. Eventually these variables create an irregularity. Figure 2b, noting the planetary positions for Sun Spot Minimum, shows that cycles 21, 22 and 23 are different. In these three cycles, Jupiter has passed the ninety-degree position. As this alignment is associated with the coming maximum, we could consider these as "irregular". The smoothed relative numbers for cycles 21 and 22 were above 12, and the average is only six. If there is an irregularity, the next cycle probably has to be irregular to put the mechanism back into its normal rhythm. These conclusions are from the study of the rhythms over the last two hundred years.

TABLE 19–PLANETARY POSITIONS (AUGUST 17 1999 AND SEPTEMBER 20 1999)

Sun, Moon & Planets	Right Ascension For Aug 17	Heliocentric position in degrees 17th	Right Ascension for Sept 20	Heliocentric positions in degrees 20th
Moon	14:26	-	20:6	-
Sun	9:45	-	11:48	-
Mercury	8:30	35	12:26	208
Venus	9:53	320	9:24	15
Earth	-	323	-	356
Mars	15:09	273	16:39	312
Jupiter	2:12	23	2:86	26
Saturn	3:01	40	3:00	41
Uranus	21:08	314	21:03	315
Neptune	20:18	303	20:15	303

The various tables help to show how the alignments are different for the sun and for the earth. The alignments for the time of the two big earthquakes in 1999 are a good example. This is particularly noticeable with the closer planets, but the more distant ones are not so different. This indicates that the slower overall effect is similar and variations are because of the closer faster planets making short-term alignments that affect the sun and the earth at different times. This means that there cannot be exactly matching graphs, in any of the areas, and overall, the study indicates that seismic and solar activity relate to each other, but not directly. In the framework indicated, this present cycle is unusual.

CHAPTER 10

A DIFFERENT MODEL FOR SUNSPOT CYCLES

Foreseeing the Sunspot Maximum

The indication from the study is that there appears to be a different and perhaps more accurate system of foreseeing the time of the Sunspot Maximum. The planets are apparently most effective in their energy output at the ninety-degree position but in terms of timing, they do not indicate the most likely time of the sunspot maximum. This in no way detracts from the effectiveness of the ninety-degree principle outlined in sunspot and seismic activity. However, it appears this is probably only part of a wider process.

This alternative method is from a further study of the positions shown in Figure 2a. It can be seen that an average of all the angles is approximately ninety degrees but the analysis overall indicates that the opposition and conjunction alignments are really the main factor in the timing. This implies that measurement should be from that alignment rather than on averages of the maximum angles. If we look at Figure 2a we can see that all the angles closing to a conjunction are past the 90-degree angle, and are near to sixty degrees from the conjunction. On the same principle the angle opening, deducted from one hundred and eighty degrees, is also near to sixty degrees. The indication from this appears to be that the "ideal" timing for Sunspot Maximum is with

Jupiter thirty degrees past the ninety-degree position and sixty degrees from the direct alignment of an opposition or a conjunction. The table below, based on Figure 2a, shows the principle of the timing in terms of the angular positions.

TABLE 20–JUPITER-SATURN ALIGNMENTS AT TIMES OF
SUNSPOT MAXIMUM

Cycle	Maximum	Degrees apart	Direction	Formula	Angle
14	1907.0	113	Opening	180-113	67
15	1917.6	68	Closing	-	68
16	1928.4	122	Opening	180-122	58
17	1937.4	69	Closing	-	69
18	1947.5	105	Opening	180-105	75
19	1957.9	60	Closing	-	60
20	1968.9	170	Opening	180-170	10
21	1980.0	20	Closing	-	20
22	1989.9	168	Opening	180-168	12

The table shows that there is a much closer average for sixty degrees. Some very low figures reduce the average and any irregularities would further distort the outcome. Nevertheless, it appears that during "ideal" rhythms the maximum would always occur when Jupiter is sixty degrees before the inline closing point. This is really one hundred and twenty degrees from the inline starting point. If we concentrate on the Jupiter-Saturn measurement, the energy output would appear to be operative between sixty and one hundred and twenty degrees. Commencing with a conjunction of Jupiter and Saturn, the ideal opening distance is sixty degrees and the ideal maximum measurement is one hundred and twenty degrees. The activity appears to depend on the actual effects from the ninety-degree angle. By using this formula, the time of the Sunspot Minimum is apparently when Jupiter reaches sixty degrees past the second inline position of an opposition. In terms of sunspots, this is the end of one cycle and the beginning of the next one. After that the planets are ninety-degrees apart and the energy is again stronger. The maximum effect ideally occurs with Jupiter thirty-degrees past the right-angle alignment. This creates a one hundred and twenty-degree angle when the angle is opening, but a sixty-degree angle when it is closing. At the conjunction or opposition, the system switches off and the polarity changes, then it repeats. In each cycle, the old spots are still showing at low latitudes as the old energy fades. Therefore, it overlaps as the new cycle starts with the change of alignments.

The positions for Jupiter and Saturn in Figures 31 and 32 show the rhythms, and further explanations follow later. They indicate how there can be variations, for example, comparisons with real positions show differences. There are obvious delays, probably caused by other planets. Computer models will perhaps be able to demonstrate how this occurs. However, there is one noticeable technical difficulty in the above model, and a discussion of the problem follows.

A Combined System

The model discussed above is so reminiscent of the generator or dynamo principle, suggested in Chapter 1, that it seems that there could be a combined system. This would include both electromagnetic forces from the planets and a generator system with the sun. As explained earlier, the principle of a dynamo is that the armature sweeps through a magnetic field created by two magnets of opposite polarity. In this case, the planets are the magnets and the armature is the magnetic field-line of the sun. In mechanical models, there are many armature loops and there would be many field lines in the sun.

When the two planets are only sixty degrees apart, the field is small and the armature can only make a short sweep, hence the low output of sunspots at the beginning. Then at the ninety-degree position, the sweep through the field is larger and probably stronger, with a better output, but the longest field is with the two "magnets" (planets) at one hundred and twenty degrees apart. This is the time of maximum output and the visual sunspot maximum. After that the system cuts out and the polarity changes. If we take the conjunction as a starting point, the armature is sweeping from Saturn to Jupiter. After the change from the opposition, it is sweeping from Jupiter to Saturn. This is apparently why the solar field and sunspot polarities change.

The problem is that after the opposition point, the first sixty-degree position is really a one hundred and twenty-degree angle from Jupiter to Saturn. It therefore seems that the one hundred and twenty-degree angle is a weak field unless earlier stimulated by the ninety-degree position. This is likely, as it is also the end-point of the previous cycle. The sixty-degree position is therefore significant, being at the end and the beginning of two consecutive cycles. However, the thirty-degree position, past the opposition, might perhaps be the actual end. If this mechanism were verifiable, scientists would be able to plot the coming cycle. The inclusion of other alignments, such as Uranus and Neptune or the lesser planets, would indicate the degree of extra activity. By applying the principle of planetary traction on the surface of the sun, they could also anticipate flare activity.

Figure 31 and the following table indicate the ideal sequence of the alignments during a double cycle. The sun is at the centre and the planets are orbiting anti-clockwise. For comparison, Saturn is diagrammatically always in the "six o'clock" position but in practical events, both planets are moving. The elliptical orbits affect the timing but tables already allow for that. From these we can perhaps foresee the main cycle and allow for modifications by the other planets. The variations caused by the elliptical orbits may perhaps be part of the reason for the "irregular" minimum alignments in Figure 2b, in which the minimum was *after* the 90-degree alignment. Alternatively, the difference may perhaps relate to he positions of Uranus and Neptune. They more exactly aligned in 1993. Either way we can see that there are never ideal regular conditions. The "ideal timetable" is therefore a tentative attempt to clarify the principles rather than offer a strict process.

TABLE 21–THEORETICAL FUTURE TIMETABLE FOR JUPITER AND SATURN

Position	Angle/degrees	Date	Stage
1	0 Conjunction	22 June 2000	Conjunction. Polarity changes
2	60 Opening	17 January 2004	End of cycle 23. New cycle 24
3	90 Opening	28 February 2006	Energy effect
4	120 Opening	2 February 2008	Maximum. Longest field
5	180 Opposition	23 February 2011	Opposition. Polarity changes
6	120 Closing	18 February 2014	End of cycle 24 New cycle 25
7	90 Closing	21 September 2015	Energy effect
8	60 Closing	9 July 2017	Maximum.
9	0 Conjunction	2 November 2020	Conjunction. Polarity changes
10	60 Opening	18 July 2023	End of cycle 25. New cycle 26

The other planets would obviously affect the dates in the table because they also have an effect. Apart from that, we can use this principle when observing the coming cycles. If the present cycle is "irregular" and as the actual Sunspot Maximum is after the conjunction position, the next cycle may also be irregular. The sunspot count usually decreases later in relation to the conjunction of Jupiter and Saturn. There was a drop in the monthly average from 132.7 in November to 84.4 in December of 1999, but there are usually some fluctuations. The following Minimum would ideally be in 2004 when Cycle 24 should commence. However, the alignments in March 2000 indicated a strong effect in terms of the ninety-degree alignment of Jupiter and Saturn to Uranus and Neptune, and this may contribute to any irregularity. As there is currently no formal support for this, it is only hypothetical. Even so, this current cycle will act as a test case because the alignments are so unusual.

The official estimate for the minimum is 2006. In any case, the average length of fall is 6.2 years but this appears to depend on how many slow moving planets are making an effective alignment. In the present cycle, the two planets likely to have an extended effect are Saturn and Uranus as Saturn takes some time to move from the ninety-degree alignment with Uranus.

Figure 32 shows the rhythms in the sunspot cycles, and demonstrates how the Minimum and the Maximum are usually regular. If we compare the lengths in time, we can see the rhythms stretching as the century progresses. Computer analysis could assist in demonstrating how this develops from the changing relationships of the slower planets, ie. Uranus and Neptune. It is these slower planets that seem to cause the delay in the Minimum. If we consider the final figures for the sunspot maximum we can see how the planetary alignments relate to the length of a sunspot cycle.

CHAPTER 11

THE SEQUENCE OF EFFECTS

Increased Volcanic Activity

As already pointed out, the different elliptical orbits of the planets modify the regularity of the planetary alignments and change the rhythm of the sunspot cycles. Figure 32 shows how the rhythm of the cycles changes and lengthens in relation to the alignments of Jupiter and Saturn. By comparing different periods with changes in the rhythm, we can evaluate the likelihood of associated effects with increased sunspot activity, for example as in 1699. Even so, it is not just a question of more sunspots, because there are intervening processes in which a sequence leads to the final effect. Nevertheless, we can examine specific effects, such as earthquakes and volcanic activity, and consider the sequence separately. As we are considering the Sunspot Maximum in this latest cycle, we can evaluate events in 1999-2000. A few details of the volcanic activity for this period will help to demonstrate the general trend in these events.

At the end of 1999, there were a number of volcanic eruptions in various locations including Nicaragua, Ecuador, Mexico and Chile. In the first six months of 2000, there were 22 significant instances of volcanic activity around the world, compared with 22 for the whole of 1999, and 16 for 1998. Mount Mayon, the volcano south of Manila in the Philippines erupted 14 times within twenty-four hours, starting on 24 February 2000, throwing clouds of ash twelve kilometres into the air. In

Iceland Mount Hekla blew its cap and provided a spectacular lava flow on 26 February. On 2 March, there were reports of renewed activity at the Pacaya volcano in Guatemala, which had erupted earlier between 23 January and 24 January 2000. The planetary alignments for that period are of special interest as they are a classic volcano pattern. The planets, including the sun and excluding Pluto, were all within a 100-degree arc, and to one side of the earth. The moon is at times outside that group of alignments, creating cross traction with some of the planets. It then moves round within the group of planets, concentrating the traction to be more effective. This pattern is of course ongoing.

The list below shows the increase in volcanic activity during 1999 and demonstrates that the volcanic activity does appear to increase as the sunspot cycle progresses. Some researchers have earlier suggested a connection of volcanic activity with sunspot activity but the mechanism has eluded them, as they did not consider the planets. The connection becomes clearer if we study the dates given below against planetary alignments on the sun and on the earth, with an appropriate allowance for the delays in the sequential processes. As a rough rule, the slower planets are more evident with volcanic activity and the sudden increase in activity appears to relate to the grouping of planets that add to tidal pressures and cause horizontal compression. Some specialists consider that there are tidal effects in the magma. If so, it would probably match into the general process. The list of dates of volcanic eruptions for 1999 is below.

January	8th; Azores, 18th; Tonga.
February	5th to 7th; Indonesia, 24th; Kamchatka.
April	19th; Shishaldin Alaska.
June	8th; Cameroon, 23rd; New Zealand, 26th; Kliuchevskoi, Russia.
July	1st; Lewotobi, Indonesia, 23rd; New Zealand, 29th; Guatemala.

August	6th & 11th; Nicaragua, 26th; Japan.
September	13th; New Zealand, 30th; Philippines.
October	17th–18th; Chile.
November	1st; Japan, 4th; Mexico, 10th; Ecuador, 20th; Nicaragua.
December	17th; Ecuador.

The update of further volcanic activity for 2000 showed a big increase, and in particular the report on Popocatepetl indicated extensive activity. A few details from the Global Volcanism Network, with summaries from the Smithsonian Institute will help to indicate the importance of the Popocatepetl eruption. The reports quoted start in 1997, as the more noticeable events showed from that year, which was one of moderate to increasing activity. By 1998, columns of ash were projected 2 miles into the air and activity was violent. In 1999, a new dome appeared, but the reports only mentioned light ash, with later comments in the year to the effect that the activity was within normal limits. In the year 2000, there was noticeable serious activity and by August, columns of ash reached a height of 3 miles, and in October, they were at 4 miles. Authorities raised the alert level and evacuated 50,000 people. The report of the eruption on December 19 said it was the largest in a thousand years.

The indication is that the alignments relating to sunspot maximum, together with a compact grouping of planets create a specific stress on the earth in terms of inner stimulation and horizontal compression with tidal pressures.

The total number of significant volcanoes for 1999 was 22. As a comparison, the table for sunspots and seismic activity in 1999 is below. The graphs for this table are in Figure 33 and the comparative graphs for yearly smoothed numbers and earthquakes are in Figure 34. Graphs and tables for 2000 are further on.

TABLE 22–MONTHLY SMOOTHED SUNSPOT NUMBERS AND
EARTHQUAKES 5.5+IN 1999

Month	*Rel Nos*	*Quakes*	*Month*	*Rel Nos*	*Quakes*
January	62	21	July	113.5	13
February	66.3	21	August	93.7	26
March	68.8	28	September	71.5	29
April	63.7	17	October	116.7	23
May	106.4	25	November	133.2	33
June	137.7	23	December	84.6	53

Flares and Ionospheric Disturbances

Extended graphs of seismic and solar activity also indicate that there
may be a connection between flares and earthquakes. In considering the
sequence, we have to include other phenomena such as ionospheric disturbances. The graphs show how the occurrence of flares follows the
main curve for sunspots. This is because most flares originate from
sunspots but a flare is a more definite event in time. As explained earlier,
many flares appear to be caused by the planetary traction on the surface
of the sun, which disturbs sunspots and helps to cause a flare.
Nevertheless, the rise and fall is slightly different despite the similarity.
With the earthquakes, the peaks appear to be later as there seems to be a
delay from the flare event to the seismic event. This is apparently
because the charged particles take time to reach the earth and there is
probably a delay in the activity within the earth. The peaks therefore do
not match exactly because of these different factors, especially as the
alignments are always different for the earth. The planetary traction
that helps to trigger an earthquake is separate, yet often parallel in

effect. However, the flare effect is not direct because it first disturbs the ionosphere.

The graphs of flare disturbances were from all listed events. Many of these were close together in time and coordinates. As previously explained, the same problem exists with earthquakes. Foreshocks and after shocks are listed as separate disturbances, although they may be close in time and have the same coordinates. For the sake of simplicity, the seismic graphs were for seismic disturbances of magnitude 4.9 and above on the Richter scale. Checks with other limits, such as from magnitude 3 and from magnitude 6 showed very similar peaks and curves. This indicated that the lists of flares could equally give a relative measurement for purposes of comparison. The analysis of disturbances of the ionosphere was therefore on the same basis. There is a difference with the disturbances of the ionosphere because the lists of disturbances only show the times for the event. These lists show the beginning, peak and ending, as with flares, but there are no coordinates for locality as the effect is more general. However, our aim is to compare sequential quantities, and for this the position of an earthquake or a flare is not important to the graphs. We can therefore consider the ionosphere disturbance lists on the same basis.

In the graphs in Figures 35 and 36, the peaks are not exactly the same, and it may be that planetary traction can disturb the ionosphere as with the surface of the earth, because other references mentioned atmospheric tidal effects. In that case, the electricity phenomena before some seismic disturbances could be from the planetary tidal effects on the ionosphere as well as from ground radiation. Irrespective of this, there is some match of the main peaks. In comparing these, the position of the quiet points also helps to assess the possible delay in effects. In general, there seems to be a delay from the flare peak to the peak in disturbances of the ionosphere. The indication is that there is some further delay before the seismic peaks following the sequence through from the flare event. The comparisons raise the question as to whether changes within the earth itself could disturb the ionosphere. If that were so, the accuracy of a sequential pattern would need some refinement. Specialists already know that there is a connection and these electromagnetic principles are the subject of current research and investigation. One reference on the Internet by the National Geophysical Data Centre said, " It is thought that there is persuasive evidence of an ionospheric precursor to large earthquakes that can be used as a predicator." This seems likely from the indications in the graphs, and the tentative suggestion is that that the important factor is the incidence of flares. That is to say more flares equates with more disturbance right through to seismic disturbances. Our final interest here is in the link with the planetary effects, both before the appearance of sunspots and before the seismic impact. On this point, it would appear that an absence of planetary alignments that create a traction disturbance of the sunspots would match a decrease in flares and the ensuing effects. Because of this, more sunspots would not exactly match more disturbances of the ionosphere or seismic activity. However, the peaks may not always be from planetary causes, as further study of magnetic storms indicates.

Magnetic Storms

We have already considered the importance of solar storms, and can now attempt to relate them to the possible sequence, as they also cause geomagnetic storms in the ionosphere. Magnetic storms are abrupt disturbances on the sun that have a stronger effect. According to the description, such storms generally occur once or twice a month, although there may be as many as five or six. The explanations say the cause is a conflict of magnetic energy involving the solar field and the earth field and there are several effects from this. The most prominent is a strong flare that would increase the normal fluctuations of the ionosphere. Another cause is from the energy in "coronal holes" around the sun. The charged particles from such holes reach a higher velocity and the significant point is that the coronal holes exist for many rotations of the sun. The effect on the earth lasts for several hours and on the day of the main impact, the energy disturbs the ionosphere more than usual. The significance of this in our investigation is that the average period of rotation of the sun is approximately 27 days. If we look at the graphs of the ionosphere, we can see that there are peaks about one month apart. We can refine the period of 27 days to a more exact time because in the 27-day period the earth has also moved 27 degrees on its orbit. Therefore, the rotating coronal hole will not directly affect the earth for another day or two. The duration of the rotation depends on the latitude and it could be more than 27 days. All this is of course well known and scientists observing the sun and the ionosphere can plot the time of impact exactly.

The importance of this concept lies in the way the graphs for disturbances of the ionosphere and seismic disturbances match. There is a possibility that the cause of a higher peak in ionospheric disturbances could be a magnetic storm, which might also match a peak in seismic activity. If that were the case, the effects from high-energy particles from a coronal hole would repeat and we could perhaps anticipate the next peak of seismic activity. The Turkey and Taiwan earthquakes were approximately one month apart, and we have the question of what helps to cause these peaks. Venus was very close to the earth with one major peak of earthquakes. The extended magnetic field of Venus from solar radiation would have been touching the earth, but it was not so close with the other earthquake and the cause of that repeated peak was not clear. Whether we can include magnetic storms here is uncertain, as it would require more research to clarify the effects.

Further to the subject of strong fluctuations in energy, there is evidence that the energy effect at the polar caps varies at certain times, as there is a greater measurement in voltage. The point is that if this energy does enter the earth it might be contributory to seismic activity. The mechanical effect of the final disturbance of the crust by external traction is therefore only a tiny part of the puzzle. The point at issue is that without all the other effects the final trigger is less effective and at times appears to have no effect. If there were a connection, the comparative observations of the ionosphere might perhaps help to clarify the problem. For example, it may be that there is an observable disturbance, or ripple effect, on the surface of the sun as it rotates into the gravitational field of the planet or planets. The same effect could occur within the ionosphere. As noted earlier there are sometimes very slight seismic movements at such times. If this effect occurs with the sun or the ionosphere, it might help to confirm the trigger effect and help in understanding how it operates.

Short Term Peaks in Sunspot Activity

The details just discussed suggest the sequential effect from sunspot activity to the flares and through the ionosphere to the earth. To start the sequence we have to include the heliocentric planetary alignments on the sun. The earlier analysis showed how Jupiter and Saturn appear to create the eleven-year sunspot cycles, with longer overlapping cycles from Uranus and Neptune. If these two pairs of planets dominate the longer cycles, it follows that the faster moving planets would relate to shorter cycles. In other words, some peaks would occur because of planets from Mars inwards, ie. the earth, Venus and Mercury. Checks on these four planets indicated that 90-degree alignments to any planet, including Pluto, did usually precede an increase in activity. Pluto was not included in the main analysis because of its small mass and distant position but it appears to be effective in the ninety-degree alignments. In general, alignments of these inner planets could add to the effect of the slower alignments or create peaks on their own. This is simple to check, as there are tables of computerised data for these positions. It is only a question of whether any alignments with these planets preceded any peaks during any month of recorded sunspot relative numbers.

The following tables of monthly peaks show how 90-degree alignments generally do precede an increase in sunspots. The first table lists
the alignments for January and February in 1998 and the later table for
January to May of 2000 shows the same type of information. There is
also a table in the Appendix for all of 1999, which relates to sunspot relative numbers for that year. These positions show the beginning of the
sequence, ie. planets affecting the sun. At the other end of the scale, ie.
with peaks in seismic activity the effect of the ninety-degree alignments
to the earth may only be secondary because the electromagnetic energy
effects are perhaps more important. The prior process to a seismic
effect apparently includes the alignments that disturb a sunspot and
help to create a flare. Presumably, if all these conditions exist, and are
contributing to inner activity within the earth, the group of external
bodies acting as a trigger will be operative.

By working from the position of the alignments, as shown below, it
seems possible to foresee a rise in sunspot numbers. However, my own
check on this was not under controlled conditions and it obviously
needs specialist observation with confirmed results for a firm statement
on this principle. It is such an easy point to check and examination of
past records will clarify how reliable this method of anticipating
sunspot activity could be. The point that continually emerges is that if
we can verify the planetary connection we could foresee many events in
solar and terrestrial effects to some extent. The tables of the peaks indicate the beginning of the sequence.

TABLE 23–SUNSPOT PEAKS AND ALIGNMENTS FOR JANUARY & FEBRUARY 1998

Year/1998 Month/No	Date of peak	Sunspot Relative no	Date of alignment	Heliocentric 90 degree alignments
January	12th	38	7th	Mars-Pluto
31. 9	..	..	7th	Venus – Mercury
	..	..	9th	Mercury – Earth
	16th	52	10th	Earth –Saturn
	..	..	12th	Venus – Saturn
			15th	Mercury – Uranus
	25th	75	23rd	Mercury – Jupiter
				Jupiter – Pluto (87)
February	14th	76	9th	Venus – Pluto
40. 3	..	..	..	Mercury – Saturn
	24th	59	22nd	Mercury – Pluto
	..	..	24th	Earth – Pluto
				Jupiter – Pluto (87)

The related graphs in Figure 35 and 36 are for sunspot numbers, flares and ionospheric disturbances during January and February 1998. These demonstrate how the general peaks in the sequence follow each other with the delays, from the sunspots through to seismic activity.

The January–May 2000 Peaks

In terms of the foregoing principles, the January to May 2000 alignments also indicate how the solar and seismic disturbances fluctuate. The table below shows the sequence of 90-degree alignments and the graph in Figure 37 for that period shows the critical peaks. The general indication appears to be that there is often a six or seven-day time lag from sunspot activity to seismic activity. At a casual glance there appears to be very little connection on such short-term graphs, yet long terms graphs seem to show similar smoothed curves. The conclusion here is that the other variables contribute to seismic activity. As noted earlier, alignments that directly affect the earth modify the effect as well as the tides. From this, we can see that direct alignments to the earth only give a broad indication of likely seismic activity. The overall process has to include sunspot activity but even this leaves some gaps. To solve the whole puzzle we need specialist knowledge of the processes in physics that would fit into the general hypothesis.

The increase in volcanic activity suggests that perhaps volcanic activity is more akin to sunspots, and the flares are more like seismic disturbances. Although the sun is a gaseous body and the earth has a solid crust the principle of an active core appears to be similar. There are so many similarities in the apparent processes, that it seems there is some connection. For example, in the following table, which is similar to one constructed for anticipating general seismic activity, it appears possible to predict variations in sunspot activity. If each planet had a radiation value, we could probably anticipate the amount of stimulation. Existing records of radio noise, magnetic field strength and so on might already contain the necessary information to confirm the principle. It also seems that we could estimate the possible sunspot relative number from experience and comparison with earlier alignments. However, the extended effects to the earth are by no means certain. For despite such a possibility there are other requirements for a terrestrial effect, as with the unique alignments in Figure 30, for May 11. This is of special interest as it is the period before Sunspot Maximum. The most critical alignments were in May, but the peak was in July because of the delay in sunspot appearances. The seismic activity was then later, with other contributory factors modifying the effects. The diagrams indicate that in general there must be suitable alignments to affect the core of the earth and a suitable group to trigger the fault. In addition, it seems there has to be an extra boost of energy from the sun. Whether this is by increased solar radiation, solar storms, or flares is not clear. The proximity of planets is also a significant factor. As this involves the ocean tides, a definite outcome is difficult to foresee. Although this is a hindrance for predictions, the extended graphs for sunspots and seismic disturbances do appear to indicate a connection.

TABLE 24–SUNSPOT PEAKS AND ALIGNMENTS FOR JANUARY TO MAY 2000

Year/2000 Month/No	*Date of alignment*	*Date of peak*	*Sunspot Relative no*	*Heliocentric 90-degree alignments*
January	8th	14th	164	Mercury – Mars
90. 1	14th	..	..	Earth – Venus
February	6th	6th	136	Earth – Saturn
112. 9	8th	16th	131	Mercury –
	10th	..	..	Mercury –
	11th	..	..	Mercury – Earth
	22nd	29th	162	Mercury – Mars
	24th	..	..	Mercury –
	25th	..	.	Mercury –
	27th	..	..	Mars – Neptune
March	1st	7th	155	Uranus-Saturn
138. 5	..	..	..	Earth-Pluto
	..	..	..	Mercury – Pluto
	22nd	24th	188	Mars – Uranus
	23rd	..	..	Venus –Jupiter
April	27th (March)	2nd	193	Venus – Saturn
125. 5	28th (March)	..	..	Venus – Mars

	19th	23rd	170	Mercury – Jupiter
	20th	..	..	Mercury – Saturn
May	6th			Mercury-Neptune
121.6	7th			Jupiter-Uranus
	8th	11th	133	Earth-Uranus
	..	12th	133	Mercury- Uranus
	13th	15th	205	Venus-Neptune
	..	20th	180	
	24th			Mercury-Jupiter
	..			Mercury-Saturn
	..			Mercury-Venus
	27th			Earth-Mercury
	28th	28th	124	Mercury-Pluto

March and May had critical alignments but the actual peak was in July.

Review

The whole process appears to rest on the principle of some type of radiation from the planets, and the indirect gravitational force that triggers the final event. For lack of an adequate terminology, the term "electromagnetic force" tentatively describes the radiation, but this description is open to debate, especially as some descriptions say that some waves cannot penetrate the lower atmosphere. From this point alone, it seems clear that electromagnetic waves would probably not penetrate the earth and therefore affect the core in that manner. This implies that the energy would perhaps have to enter at the poles, or is a completely different type of energy.

More than one scientist has commented on the fact that the formula for a gravitational effect is the same as that used with magnetic force. Furthermore, it seems that there is an effect on the ionosphere, which could be from gravitational force or an electromagnetic effect. Overall, it seems wiser to suggest that the planets do radiate some type of electromagnetic waves. In the specific combinations, there are some noticeable effects that scientists could certainly observe and measure. This applies to both types of effect, ie. on the core and on the surface. Either way, the analysis of diagrams can only go so far, and if there is an input of solar energy at the poles of the earth, diagrams will not show this. Specialist field study could record the measurement of increased polar lights or energy input that may precede seismic disturbance. The clarification of these conjectures therefore rests with the experts, and it is to them that we must look for specific explanations.

There are some exceptions to the simplified outline, as there are times when the conditions seem to apply but there is no effect. Whichever way we study the problem, it appears to start and end with a planetary effect. In considering this, it may be relevant to mention the areas that made news headlines in terms of human disturbance during the month of May 2000. They were Sierra Leone, Israel/Palestine, Eritrea/Ethiopia, Indonesia, Philippines, Solomon Islands, Fiji, Zambia, Nigeria, Peru, Ecuador, Sri Lanka, and Venezuela. Some of these are in seismic areas but in any case, it would be interesting to know variations in ionisation and magnetic fields that might have helped to disturb humans. Another news report mentioned an intriguing case of a young man who was noticeably disturbed at the times of Full Moon. The report stated that the doctors were baffled and unable to account for the psychological disturbance. Two interesting points mat be relevant in this case. One is the extra ionisation of the earth's field at the times of Full Moon. The other relates to animal disturbance before earthquakes. Scientists now know that animals are disturbed because of the extra positive ions released from the rocks under pressure. This developed from the astute observation of a scientist who was in a railway yard at the time of an earthquake in the area. He noticed that although animals were restless that cattle in the metal railway trucks were not disturbed and concluded that this was because the metal trucks acted like a Faraday cage and screened them. The implied solution to the distress of the young man might therefore be in the use of a type of Faraday cage. This would probably be more effective at night, when the moon is above the horizon and reflecting solar energy into the earth's field. If correct, it would be further evidence of human distress from celestial objects and an indication that the true explanations are in biophysics. In view of this, it seems appropriate to examine the ancient science of priestly astronomy practiced by the early naked eye astronomers.

CHAPTER 12

THE SCIENCE OF ANCIENT ASTRONOMY

Do Planets Influence People?

The ancient science of priestly astronomy and naked eye observation developed into a system of associated effects based on coincidences. This has developed into the subject of astrology, where coincidental planetary alignments are an arbitrary indication of impending events. In this, the question of scientific principles was secondary to the fascination of trying to foresee the future. Yet, we have to ask, is there a basis to the assumption that planetary alignments affect life on earth? Without stretching this point, it may be that the principles discussed earlier indicate how astrology might operate. From this we can consider the possibility of the influence of planetary bodies on people and their behaviour. Practitioners of astrology claim that the planets influence people, and since the planets are physical objects operating within the framework of natural laws, the clear indication is that the subject should be within the laws of physics. Furthermore, people are biological organisms so any effects would have to be within the framework of biology. Together they indicate that there should be explanations in terms of biophysics. This is the standpoint in this present discussion and it appears demonstrable that ancient astrology actually did operate within a rational framework. However, the earlier priest-astronomers did not understand physics as modern scientists know it.

How does it work?

The discussion here deals with the further secondary effect from the planetary forces that act on the core of the earth and on the surface of the earth. The main thesis is that these effects modify the field of the earth and thereby affect biological life in that field. As this includes human beings, the indication is that such changes in the earth's field do affect people, the mechanics of which were covered in an earlier chapter (Chapter 7) on biological effects from electromagnetic forces. The approach here is their application in terms of classical astrological principles. This approach does not consider the modern expression of popular astrology but discusses the natural laws that led to the development of what we now describe as astrology. The aim is to show that natural forces operating via the planets can affect people; in technical terms, this would be "astro-biophysics". It is demonstrable that all the propositions are in the range of formal science. The principles discussed are therefore not only within a framework of stringent observation, but they are capable of duplication under laboratory conditions. However, this does not mean that reactions or responses are inevitable, but only that these influences do exist. How we react to natural conditions is a matter of awareness and personal responsibility. The standpoint taken is that these natural forces do not create conditions or events, but only affect people via their nervous system. Consequent actions are then a product of choice, rational or otherwise. In this sense, it may seem that events appear spontaneously and that we are not making a choice. This debatable question lies outside this inquiry as we are here only considering what the forces are and how they operate.

Moon Madness

The effect from lunar forces has often interested investigators and one of the most intriguing reports was in a work entitled *The Lunar Effect*, by Dr Lieber, which we have already discussed. As pointed out, his investigation did not pursue the effect of ionisation caused by horizontal compression, although there was a reference to the coincidence of high tides at the time of disturbances. We have already considered how the earth field changes at the times of Full Moon, but the subject of ionisation from tidal forces needs further explanation.

In seismic studies of animal disturbance before earthquakes, the principle is that positive ions disturbing the animals are from horizontal compression of the tectonic plates. The hypothesis put forward here is that some horizontal compression occurs with every high tide, and when the planets add to the effect of the sun and moon there is extra horizontal compression. This releases more ionised particles with a consequent greater disturbance of the biological life. Observers watching the movements of the celestial bodies would note the coincidence of the external bodies that matched the restless behaviour of people. This could then become a "rule". From this, and the effect of the Full Moon reflecting solar radiation into the earth field, there would be two such rules. Both would rest on the effect of a more ionised field and as various planetary effects could modify the field, a need for a wider set of rules became evident. These arbitrary rules were from observation of coincidences, and not from knowledge of the actual process.

The Planetary Effect

The main influence of the planets appears to occur in specific alignments. This may be via two or more planets in a ninety-degree alignment, as explained earlier with solar and seismic activity. The proposition put forward was that the planets radiate electromagnetic force from core activity, and when combined there is an extended effect.

This extended effect is where the electromagnetic radiation from two planets at 90 degrees apart intersects. This means that if the earth is at the intersection, the core of the earth is stimulated, and increased activity of the core modifies the magnetic field of the earth. If there is also an ionisation effect from the horizontal compression, as already explained, or from the Full Moon, or from solar flares, there may be a stronger and more ionised field. The biological life in that field is then affected, and some organisms experience more effects than others. This last point is important, and we will look into its significance later.

Solar Flares

We have already considered how solar flares increase with sunspot activity and in addition, that seismic activity increases with solar activity. This means that there is likely to be an increased incidence of animal (and human) disturbance as the sun spot cycle progresses. As the sunspot cycle relates to the planetary alignments, we can again point to an influence via the planets. However, there is not just one direct effect from solar flares. The flares can affect the earth field directly and thereby add to the ionisation of the earth field. In addition, they can cause a fall out of charged particles from the ionosphere, and they may stimulate the core of the earth by entry at the poles. This shows by the polar lights that fluctuate in intensity with solar activity. From this, we can deduce that the welfare of the earth changes with the planetary effects on the sun, as well as on the earth. As these effects are specifically within the framework of physical laws, they are open to checking and verification, and could further support the concept of effects on biological life.

The critical point is that the planetary radiation affects the core of the earth directly as well as the extended effect via the sun. From the combination of electromagnetic radiation and ionisation from the sun, and by horizontal compression, we have a definite framework of planetary influences. As tidal forces from external bodies cause the horizontal compression, either by tides or directly, as earth tides, we almost have a complete frame of reference for explaining planetary effects on biological life.

The Human Connection

From the foregoing details, we can see there are specific effects that could be open to routine experiments and ordered observation. What we still need is a process of how these influences specifically affect individuals. The previous explanations only indicated a general effect, but astrology claims specific effects. Traditionally a birth chart shows how one individual is uniquely affected. Again the essential question is, how does it work? The operative point, which we now understand from laboratory work, is the principle whereby any greater field affects the lesser field within it. This may be a type of magnetic resonance, yet it does not appear to happen with every human bio-field because they are not all the same. The work described in *Psychic Discoveries Behind the Iron Curtain* (Ostrander and Schroeder 1973) said that researchers of Kirlian photography noted that biological life forms did not have an exactly similar field and that there were observable differences. They also said that there were noticeable effects in the biofield at the times of solar flares. We considered some of these points in the chapter on biological disturbances. Although it appears that mutually similar fields would very likely lead to mutually similar reactions, the indication is that the individual bio-fields are different. This would be the reason why there are subtle individual effects. We therefore have to consider the personal field and ask how it could be different from another personal field.

The Individual Bio-field

Astrologers insist that the birth time is significant. Scientists have generally dismissed this and so far, astrologers have been unable to refute the dismissal. Yet, the clue appears to be in the positioning of the planets that affect the earth field. This is because the changing alignments would affect both the sun and the earth. By this process, the earth field is continually changing. These changes in the earth's field would cause different effects on the biological organisms living within it. Although they are subtle, these differences would occur because each individual has a slightly different field and would therefore react differently. The question then arises, how and why does an individual have a unique personal field? The answer to that appears very simple, although it is probably complex. The critical point is that before birth the infant is in the womb and shares the mother's bio-field. At birth, when the infant escapes from the mother's field it acquires a personal field. Where does it obtain that field? The indication is *that the child's individual field forms from the prevailing earth field.* The new individual then has an "imprint" of the magnetic field of the earth at that point in time, at that place. As the earth field is not the same in all places and is not the same all the time, the positioning and the timing would be significant. The details would then indicate a formula for the individual field. Of course there are other contributing factors, such as hereditary and environmental influences. That is to say, via the genetic, social and educational forces. Even so, there is an underlying framework of basic characteristics that relate to the existing electromagnetic superstructure.

Can We Test It?

As the basis of these conclusions is within physics, it follows that verification or refutation of the propositions is possible. One possible test is to check the earth field at the times of critical alignments and ascertain whether there are repeated changes and effects. Another is to check the radio noise from different planets at specific times. These times would be when a planet is at the focus of other planets in critical alignments. The indication is that the same process applies to all the planets and the radio noise would vary with effects on the core of all those planets. However, these technical astronomical and geophysical methods need a high level of expertise and equipment. Nevertheless, there are laboratory methods as well and one that seems worth examination would be with mice or other small animals.

The advantage here is that the radiation from the planets and the strength of the earth field is quite low. Exposure of animals, or even people, to such influences would therefore be within the normal range. The test would be to simulate the radiation from two planets, say Jupiter and Saturn, and observe the effect at the focus of a ninety-degree alignment. Instrument readings at the focus would show whether there was a significant effect and mice in a cage at that point would hypothetically be disturbed or restless when the radiation made suitable intersections. The three main areas for examination would probably be in variations of the electromagnetic waves, magnetic fields and ionisation, within the normal ranges. Testing outside the normal ranges would produce an artificial result. Laboratory instrument tests might even suggest what happens in the core of the sun. That is, when nuclear activity is stimulated by such microwave radiation.

From a biological point of view, testing is more feasible in a laboratory. The four main points in the earlier chapter on biological disturbances indicate why.

- Each individual has a different personal field.
- All lesser fields tend to resonate at the frequency of the larger field but the individual differences modify the effect.
- Dominant tendencies increase by electromagnetic stimulation.
- An effect can persist after the initial stimulation.

Research within this framework would initially need to concentrate on the biophysical aspects of various scenarios, to establish and measure links between individual experiences and external electromagnetic stimulation. Scientists are already researching possible negative effects from mobile phones, but "astro-biophysics" would require a broader study of natural fields.

Classical Astrology

According to histories of the subject, the ancients did not include Uranus and Neptune and only worked with outer planets as far as Saturn. With these and the inner planets, they had various standard rules, which they used to evaluate the situation. These rules of interpretation rested on observation of the alignments and positions, which modern texts describe as "aspects". In general, they were either good or bad, and most writings carry on this tradition. The bad aspects were the conjunction, the opposition and the ninety-degree alignment, called a "square". Starting with the square, we can consider these alignments in terms of electromagnetic waves. The square appears to be the most disruptive and observation probably confirmed that it preceded various disasters and disturbances. Without need of knowing the exact processes, a right-angle alignment was bad. There were six or seven such planetary possibilities, and the sun and the moon at the Neap Tide probably added to the conclusion that the ninety-degree alignments did not bode well.

The conjunction, of the sun and moon, or planets in the same alignment stood out as dubious configurations, especially if the rising and setting coincided with an earthquake. The opposition was also a poor alignment and events at the Full Moon added to the conclusion that it disturbed susceptible people and in extreme cases contributed to common "lunacy". As explained earlier, under certain conditions violent people tend to be more violent and disturbed people are more disturbed. Ignorance of the cause, from ionisation or magnetic changes, did not prevent observers from noticing the effect. With planets, the opposition of a major planet with the sun meant that the planet was at its nearest to the earth and probably coincided with seismic activity if other contributing factors were operating. If some of these alignments coincided with unusual weather, they would be associated with floods or drought, and over the years a whole spectrum of possible effects would eventuate. The "good" aspects were probably only the ones that were not "bad". However, these observations did not consider any processes in the sun, which in this scheme was only another celestial body that could be in a good or bad position. Overall, the rules were very general and did not explain that a "bad aspect" may depend on specific conditions to create an effect. The broad rule was that more "bad aspects" of any type, meant more doom and disaster, and in general, this was a reasonably adequate theory.

Planets that were rising were significant, and in terms of a trigger for earthquakes they certainly were. Overall, the early naked eye observers had a practical system that made sense even if the laws of gravity and modern physics were a mystery. Unfortunately, the subject degenerated into a market place activity with lists of matching phenomena that gave a ready-made interpretation. Consequently, some modern practitioners assume that the "aspects" create the events, rather than recognising that the increased energy only stimulates action. Behavioural psychologists may see it as another science and perhaps it really is the "physics of psychology".

The Delphic Effect

This reference is of special interest in relation to earthquakes, and is an interesting example of how a coincidence appears mysterious when specific connections are unknown. The Oracle of Delphi was a priestess who acted as a medium or channel and received guidance from the Greek god Apollo. One recorded account said, "When an earthquake occurs a king will fall." We obviously wonder where the connection is. Yet, in terms of the foregoing principles we can see an explanation. The earthquake would relate to a release of positive ions and electromagnetic effects. There would be human disturbance before (and after) the earthquake. The existing dominant tendencies would then increase. A resentful populace would become more resentful and the king might then fall. The king's astrologer would probably help to avoid this by advising the king when to double the guards. This may seem whimsical but many of the world's trouble spots are on or near major tectonic intersections where plates meet. These locations would experience a greater release of positive ions, and scientists undoubtedly know this, but it has not so far been associated with human disturbance.

The essential point for consideration here is that the natural forces do *not* create circumstances. They only stimulate or disturb the individuals who then act on their prevailing moods and attitudes. Some would react blindly and subjectively, but more objectively aware persons would not. In the last analysis, the individual is self-responsible because these forces and influences have always existed, and are a normal part of our wider environment. In considering this, it is of interest to review various world events in this framework. Particular regions that stand out as areas of human disturbance are the Balkans and Indonesia, as well as the Middle East. The events raised the question as to whether they may be an example of human disturbance before earthquakes. One researcher, working from a different approach, went to the Balkans to collect rock samples, saying that magnetic changes affected the molecular structure of the rock. He also said that these magnetic changes affected people, but did not say what caused the changes.

In view of the human disturbance in the Balkans and Indonesia, the records of seismic activity had a special interest in relation to the developments. There was one earlier model, in the fall of the Sha of Iran. This started with human restlessness and there was a big earthquake near Tabas in September of 1978, followed by many big after shocks and great social changes. The Sha left Iran in January 1979 and the new government eventually took over. In the areas under consideration, the human factor has become all too obvious, and we can easily jump to conclusions. However, the interest was in the possible human disturbance as a precursor of seismic activity. In 1999, there were two serious earthquakes in Eastern Europe and the Far East. One was the Turkey earthquake and the other was in Taiwan. These earthquakes were in areas *near* where there was social unrest, which is not specific enough. In the Middle East, where three plates meet, there is unrest but no earthquakes, although there may be more ionisation.

Early Observers

In searching for indications as to how early astronomers viewed the subject of planetary influences, one or two names stand out. The first is Claudius Ptolemy who lived in the first half of the second century AD. He was a native of Egypt and lived in Alexandria. His main work was a summary of Greek astronomy. The central theme was that the earth is a sphere at the centre of the universe and the planets moved round it in circles. With this system the astronomers were able to plot the movements of the planets and allowed for the apparent retrograde motions by various calculations. The scientific discoveries of Copernicus and Kepler in the 15th 16th and 17th centuries eventually replaced the Greek concept. The interesting point about Ptolemy was that he wrote a complex book on astrology called *Tetrabiblos* and modern astrology developed from this work. An English translation is available and any astrologer would still find it understandable.

The German astronomer, Johann Kepler, 1571-1630 took a serious interest in astrology and is said to have advised astronomers and scientists not to "throw the baby out with the bath water". He apparently considered that astrology contained some useful knowledge, and some of his letters refer to it. Another famous astronomer who studied astrology was Tycho Brahe, 1571-1630. He rejected the heliocentric model of the solar system proposed by Copernicus, 1473-1543, and said that all planets revolved round the sun, but that system revolved round the earth. Sir Isaac Newton, 1642 –1727 also examined the subject seriously and did not dismiss it. Histories of astronomy are careful to ignore such details and only make allusions to Newton, and others, having "strange interests". By all accounts, astrology is indeed the Cinderella of the sciences, and scientists continue to dismiss it as a pseudo science.

There are no serious explanations as to why the early astronomers were priests and the general view is that they needed to know when to plant their crops. However, the eastern teachings indicate that the priest-astronomers were concerned with foreseeing very long cycles in time in which the influences from the wider environment affected human welfare. Western science does not favour this cyclical view and seems to prefer a linear development that started with the "big bang."

It is easy enough to assume that man is unique in this overall process and that his consciousness is independent of any external influences. Modern biological philosophy takes the view that a person's condition is a result of various chemical states that affect the brain, and therefore behaviour. That is to say that our attitudes and behaviour are a product of biochemistry. However, the indication is that there is an equal effect in terms of biophysics. It is very simplistic to suggest that all biological effects are due to charged particles, magnetic fields and electromagnetic forces, but the fact remains that we do live in a huge magnetic field that is continually being disturbed and modified by other forces and energies. Perhaps some specie declined because of such changes and from this we could speculate that we too could change because of external influences on our biological force field climate.

External Effects

An example of external effects is from the Gamma rays and X-rays, which are still bombarding us from the Crab Nebula, first sighted in the 11th century. The first sighting was on July 4 1054, and as it is about 5000 light years away, the visual light only became visible at that time, although the explosion occurred much earlier. The charged particles, known as rays, take longer to reach us. For example, the light from the sun takes 8 minutes to reach the earth but the X-rays and other ions can take days. From this knowledge, we could probably calculate when the influx of harmful rays from the Crab Nebula reached the earth. We might have difficulty in relating such effects to biological life but if we could it would, technically speaking, be applied astrology. The situation is that a demonstrable effect from any extra terrestrial influences, whether by charged particles or electromagnetic forces, would open the door to a wider investigation. Scientists admit that radiation and microwaves can have a serious adverse effect biologically. What they do not yet concede is that radiation from outer space might have a subtle personal effect. In this, there might be no observable biological change. Yet, there would be an effect through the nervous system, and this would show as psychological effects. As earlier indicated this would stimulate the existing psychological attitude and might show as an aggravated effect or a benign effect. The operative point is that existing psychological attitudes are stimulated. That is, the incoming forces do not create them but only amplify them.

Researchers are already working in this field, presently called astro-biology, although it should really be astro-biophysics. Irrespective of the name, we can no doubt look forward to further discoveries in the subject.

CHAPTER 13

PREDICTIONS FOR THE NEXT CYCLE

A Review of the Generator Hypothesis

A statement in the NASA astronomy section of the Marshal Space Flight Centre Website (April 13, 1998) said that scientists believe that "sunspots are driven by the dynamo that lies hidden beneath the photosphere, but they are unsure of what controls the dynamo."

Dr David Hathaway, in the same report, is of the opinion that no real physics are involved and predictions are made by statistical inference. However, the statistical details presented here suggest that there is a process in physics, because the indication is that *the planets, especially Jupiter and Saturn, drive the dynamo*. The concept of the solar physicists is apparently of an internal dynamo with no external stimulation, as suggested in my mechanical model. Another NASA Webpage[1] explains the conventional view of an electrically conducting fluid, which moves through a magnetic field or plasma, such as regions of ionised gas. Even so, their model operates on motion because the fluid system is a moving system. They add that one earlier view was that the cause of sunspot magnetic fields was also from a fluid dynamo system, although it was apparently still a self-contained system.

---

[1] David P. Stern & Mauricio Peredo, Electric Currents from Space/ The Exploration of the Earth's Magnetosphere. (NASA Educational & Goddard Space Flight Center)

Their details further indicate that the same applies to the earth, and in general, scientists consider that a fluid dynamo in the core generates the earth's magnetic field. In their models, there is not an external supporting system, as proposed here. One scientific explanation was that the extended tail of the earth's magnetic field could in effect act as an energy collector, in which ionised fields moved through the tail, or as the earth moved on its orbit. In the planetary model, the field of the sun helps the sun to absorb energy. The essential difference in my model is that the planetary process, depends on an ordered repetitive system with optimum points of energy intake. The present established concepts seem to be of self-contained systems, even though they speak of a sun-earth connection. From their accounts, this is an interaction of space plasma with electromagnetic fields, in which the fields are interplanetary or localised on the sun. A strict connection with the planets is not necessarily part of their models. Nevertheless, their models rest on known principles in astrophysics, while my model may appear too mechanical.

Despite that, the planetary model implies that the planets are both collectors and senders of energy in an integrated generator system *where they all act on each other*, as they do with gravitational forces. That is, they are actively absorbing and radiating an electromagnetic energy and are not just moving magnets as in a mechanical generator. In this complex planetary generator, the sun's activity is the most noticeable because of the sunspot cycles. This study also shows that we have seismic cycles, which implies that other planets perhaps have cycles of some type. The repetitive statistics on the planetary process clearly show that the gradual rise to Sunspot Maximum relates to the major planets making the heliocentric ninety-degree alignments. These match phases in the activity of the dynamo principle in the solar generator. The interpretation from this is that the ideal eleven-year planetary configurations are with Jupiter and Saturn as the main operative pair of planets and Uranus and Neptune as another key pair on a longer cycle. For example, the maximum in 1957.9, which had a high smoothed monthly mean of 201.3 and was the highest ever recorded, was when Uranus and Neptune were 81-degrees apart[2], with Jupiter and Saturn 90 degrees apart in February 1956 and moving to 60-degrees apart in December 1957. In addition, Mars was 90-degrees to Saturn, Venus was also 90-degrees to Saturn and Earth was 90-degrees to Uranus. Altogether there were four strong 90-degree alignments with Jupiter and Saturn 30 degrees past the 90-degree position. This was quite a strong pattern, although not ideal, as truly ideal arrangements seldom exist. (The ideal pattern is in Figure 3).

From observation, the practical indication is that although there are the two key alignments any two of the slower planets could act as the main drive in the dynamo cycle. This present sunspot cycle, No 23, is a good example, in which we had the recent combination where Jupiter and Saturn together made 90-degree alignments to both Neptune and Uranus and this pattern is uniquely different from the 1957 positions. However, this did not involve the two pairs of key planets in the ideal relationship, ie. Jupiter/Saturn as one pair and Uranus/Neptune as the other. This is perhaps why this recent maximum was not as high in sunspots as the 1957 maximum. A further point of interest is that the 1957 maximum had a "left hand pattern" and comparisons, in Figure 2a, show that this arrangement broadly matches higher sunspot numbers. In other words, a cycle from a starting point with Jupiter and Saturn opposite appears to be stronger than a cycle that starts with Jupiter and Saturn in a conjunction position, although other alignments also have an effect.

In any arrangement of these slower major planets, the lesser planets continually make combinations, which are always adding to the main trend. When this occurs, the effect increases. The analysis indicates that

---

[2] They were 90 degrees apart on 29 April 1955

when these lesser planets make a 90-degree alignment, the short-term peaks increase. When major planets move into a 90-degree alignment the main effect occurs and eventually subsides when they move out of it until the sunspot cycle ends. However, the lesser planets continue to create the general short-term peaks, even during the minimum period.

The Sunspot Maximum

My own calculations for the peak of this cycle were on that hypothesis, and from this the indication was that the best effective combinations would be in March to May 2000. Hypothetically, there would be a delayed effect from that period and the sunspot activity would eventually decrease after that, especially as the conjunction of Jupiter and Saturn occurred in June and maximum activity is usually *before* the conjunction. Records showed that the sunspot activity continued to increase *after* the conjunction in June. This further seemed to indicate, as already discussed, that this cycle is an "irregular" one. Irrespective of the anomalies, the overall alignments of the major planets would still relate to the maximum. In connection with this, the table of heliocentric positions in the Appendix gives the dates of the main alignments. This shows when the major planets make an effective alignment and the critical question is how much delay there might be in the short or long-term effects. With the main cycles, the rise to maximum is shorter than the fall to minimum. The average ratio is 4.8:6.2 but there are considerable variations in the actual figures. In cycle No 12, when the maximum was in 1883.9, which appears to be an irregular period, the rise was 5.0 years and the fall was 5.7 years. This made a near average cycle of 10.7 years. The planetary model suggests that this present cycle would have to be irregular in those terms and be a very short cycle of eight to nine years. This would be from minimum in 1996 to minimum in 2004, with the rise and fall almost equal. Working on statistical averages and assuming a regular ten to eleven year cycle, the official anticipated date of minimum would be in 2006, but there have been some very short cycles. This raises the question as to how far averages are reliable and whether the planetary model is accurate.

In the Appendix, alignments with Mercury are not included as there are so many. This is because Mercury completes one orbit in 88 days and makes a 90-degree alignment to the other eight planets twice during every orbit. Mercury appears to contribute to the effect with very short-term increases in sunspot activity. Alignments to Pluto are not in the list. The graphs show that the main rise relates to the major slower planets and the various peaks relate to the other lesser planets, ie. Mars, Earth and Venus, making alignments to each other and to the major planets. With Mercury, there may be a quick series of 90-degree alignments if other planets are in a suitable position. In which case there is an extended rise until Mercury moves out of an effective alignment, as for example with the May/June 2000 peaks. Overall, the short-term peaks are more noticeable with the faster planets, especially Mars, Earth and Venus.

Graphs from 1998 to 2000

Comparison of various graphs, such as Figures 37, 39 and 41, shows matches in the rise in sunspot, seismic and volcanic activity. There is of course the delay from solar activity to seismic/volcanic activity. The general rise is not so noticeable in the first part of those graphs, but the extra activity usually stands out. Such sudden rises are often useful for checking the main hypothesis and an analysis of the positions of the planets in this period is of some interest. A check on the heliocentric and geocentric positions generally shows the reasons for the differences and why sharp rises occur. Comparisons with other periods show a trend of slow rises with the slower planets and sudden rises with the faster planets. Yet, within this similarity there are always unusual peaks, especially with seismic activity, that indicate other causes. Even so, these appear to be within the planetary framework and we can therefore use planetary positions as a datum line.

In general, the effect of the alignments appears to be within 10 degrees of the ninety- degree optimum position. For the sun, the major planets may be outside those limits but there seems to be a more noticeable effect when they are within plus or minus ten degrees of the ninety-degree alignment. For the earth, a similar situation seems to exist within those limits, but there is the added effect from the sun. There may be a short period with more seismic activity even though earth alignments are not very prominent. Allowing for this, the alignments seem to match the different rises is seismic activity in the various graphs. Checks with other peaks show that 90-degree alignments regularly coincide with an increase in sunspot activity. Some examples are below and they indicate that the repetitive coincidences must relate to a definite process in physics, and this appears to be the planetary dynamo effect.

Heliocentric Alignments and Sunspot Activity

The graph of the sunspot activity in Figure 39 shows a very definite rise as the Saturn-Neptune 90-degree alignment begins to come into force in 1999. The sharp rise in July 2000 appears to relate to the short-term medium strength 90-degree alignments in that period. The very noticeable peaks help to match with alignments. Saturn and Neptune were 80 degrees apart on December 1998, as shown in the table. The other major alignments were then forming and by all indications continued to create the rise in sunspot activity. Figure 38 is another example of this.

The rise in sunspot numbers in July and August of 1998 appears to be the result of a series of Venus alignments. They were, July 7th Venus/Neptune, 13th Venus/Uranus, 19th Earth/Saturn, and in August, 6th Venus/Jupiter, and 30th Venus/Saturn. These last combinations of Venus, Jupiter and Saturn, are with planets that apparently have a strong radio noise. This suggests that their radiation effect is stronger and that perhaps such alignments are more effective. In 1999, the sharp rise again matched similar alignments. These were, July, 13th Earth/Jupiter, 29th Venus/Saturn and for August 1999, 3rd Earth/Saturn, and 10th Venus/Saturn. The indication from this is that the earth has a strong radiation because it has an active core and volcanic activity. The earth would therefore appear to be an effective planet in this scheme of interpretation. Conversely, this implies that if we monitor the noise from the earth we could anticipate when the earth is becoming more active in the interior. This could be a warning factor. The same applies to the sun in that an increase in radio noise from the sun precedes an increase in sunspots. As the internal processes appear to be similar, we could perhaps learn from that when seismic activity is imminent. In any case, analysis of the graphs, which give a match of alignments with rises in the graph, shows that alignments could indicate a likely rise in sunspot activity. The prediction of seismic activity is much more difficult, because there are other factors. With the sun, the planetary effect seems more direct and prediction appears to be much more likely.

Monthly Heliocentric Alignments

The graph of monthly alignments in Figure 39 gives a comparison with monthly sunspot and earthquake figures from 1998 to 2000. As explained with Table 2, the graph helps to clarify the effect of the 90-degree alignments in each month. The indication is that different combinations have a different degree of effect but the graph only considers quantity, as a qualitative evaluation would need knowledge of the specific radiation strength. There is of course no long-term rise or fall in the alignment graph as there is with sunspots and earthquakes. The peaks only indicate that there are more alignments and therefore a likely rise in the sunspot activity, and on a daily graph, they indicate a rise in sunspots within a relatively short time. From a study of the monthly graphs, an extended delay appears to operate. The indication is that there is a quick boost followed by a longer rise because the effect persists after the alignment. The process appears to be that although the power switches off, "the pot continues to boil" and sunspots still appear on the surface. The analogy with the earth is that this would lead to continued seismic and volcanic activity. Any combination with slower planets apparently gives a more drawn out effect, but any combination with the faster planets is shorter and relative to the orbital velocity of the faster of the two planets making the 90-degree alignment.

As we now have reliable sunspot records for the past 150 years and accurate tables of planetary positions, verification of this connection is simple. Overall, the confirmation of a planetary effect is easier with sunspots than with earthquakes. This is because the earth is subject to a wider range of effects, and tremor strength can vary, but the planetary connection is still calculable and within practical scientific investigation. Figure 39 shows the matching peaks from alignments to sunspots and then to earthquakes. The major 90-degree alignments (see Appendix) help to show how the alignments relate to the peaks in sunspot activity when plotted onto the graphs. The graph of seismic activity only relates to the sunspot activity in the broad effects as variations in seismic activity apparently depend on other factors. There are direct alignments to the earth, effects from solar flares, effects from the ionosphere, varying effects of the sun and the moon, such as proximity and in many cases, the effect of ocean tides, and planetary traction. These other effects are therefore extra. Despite these complications, the main alignments and sunspot activity are an indication of likely seismic activity. Worldwide phases of seismic activity are relatively easy to foresee by these methods, but specific predictions would need local expert knowledge.

Other Heliocentric Alignments

The study mainly considered the 90-degree alignments, but some others appear to be an extension of the same principle. In the dynamo process the 60-degree alignment, although probably weaker, apparently starts the process, then, after the 90-degree alignment the progress to the 120-degree positions continues the sunspot process. With major planets, the extended activity is presumably stronger, whereby there is a longer after-effect and the same would probably apply to the earth. That is to say, a similar process causes extended seismic activity and major 90-degree alignments would therefore indicate a spate of earthquakes in the same way that there is a spate of sunspots.

These conclusions are from observation and the consideration of the principle of a generator with weak and strong magnets. In this, maximum efficiency is with the strongest magnets at the optimum distance apart. Stronger magnets would create a larger overlapping field with a longer after effect. However, the planets do not really have magnetic fields in the mechanical sense. Yet, in principle the electromagnetic radiation appears to create fields that can overlap in a similar manner. The simple concept is of two radiation forces creating a different effect according to the way they overlap. The point at the focus then receives a stronger or weaker energy impact. Whether it is the sun, or the earth or another planet makes no difference. The impact is relative to the type and strength of the energy and the way they combine. In this process, distance is apparently not a factor. Theoretically the principle of such a "radiation generator" could be rationalised using a working model.

The effect is not instantaneous and the arrangement of the magnets or energy sources appears to modify the effects whereby the definition of a fixed process is difficult to detect. These variations appear to create differences although they are part of an integral master process, but for purposes of discussion we speak of averages and "normal" cycles and consider the functions separately.

The Sunspot Maximum and Irregularities

From the graphs, we can see that July 2000 is the month of highest sunspot numbers, and the date of the highest relative daily number is July 19th. On that date, Jupiter was one degree past an exact alignment with Saturn, which was at 53 degrees in heliocentric coordinates. In terms of the alignments in Figure 2a, this appears to be an irregular position as the maximum is usually *before* these two planets have reached the alignment of a conjunction or an opposition. In this model, the alignments in May 2000 could have created the high peak in July. In May, there were twelve exact heliocentric 90-degree alignments. This included Jupiter at exactly 90-degrees to Uranus, and Saturn was 93-degrees to Uranus and therefore apparently effective. In May, Jupiter and Saturn were still two degrees from an exact conjunction. If there is a delay from the alignments to the actual effects, as the graphs appear to suggest, we could argue that in terms of the operative positions of Jupiter and Saturn, this is still a regular cycle. This is because technically the maximum number of 90-degreee alignments occurred before the exact conjunction of these two major key planets.

If we disregard the faster planets, we could also say that the critical alignments of Jupiter and Saturn to Uranus were before the conjunction, because Saturn made the 90-degree alignment to Uranus at the end of January. The exact 90-degree alignment of Jupiter and Uranus was in May and the other alignments would increase the effect. The argument about a possible irregularity therefore depends on whether we take the actual mechanical positions of the planets as significant or match the positions against sunspot numbers. The original analysis was from the statistical positions of the planets in relation to sunspot numbers. Yet, if the dynamo process can be verified the technicalities in physics would determine the cut off points. The crux of this appears to depend on whether a regular delay from alignments to visible sunspots is a normal part of the process. In view of this, the question as to whether this is an irregular cycle may be debatable and in the end there is no irregularity, because it is all within a natural process.

Different Predictions

The prediction by NASA/Marshall scientists of the Space Flight Centre (1998) puts the peak of the curve for maximum sunspot activity in the first half of the year 2000 and states that "the peak interval starts in 1999". This statement closely matches the planetary model and maximum in 1999 or early 2000 would still have fallen within the range of a regular planetary pattern. My approximation for the maximum by the planetary model was March or May 2000. This did not adequately allow for an extended effect and the consequent delay. Apart from that, the graphs show that sunspot numbers increased into the second half of the year 2000. The NASA graph also projects the decline to the minimum in the year 2006. Controversially, the theoretical model based on planetary alignments gives a date in the year 2004. As already noted, the NASA Internet report of April 13, 1998 predicted the sunspot activity would peak in 1999/2000 with a sunspot count of around 170. Later, an IPS Space Services report dated October 3, 2000 put the time of Maximum in December 2000 with an average peak figure of 140. There is also a graph, which gives the peak at the end of the year and the final decline to the end of 2006. Their methods may perhaps include a consideration of planetary fields, although the report indicated that estimations were from comparative statistical analysis. The report gave predicted solar flux values to the end of 2009. These are by a panel of experts and the report said, "Predicted values beyond the cycle 23/24 minimum are made on the assumption that cycle 24 will be similar to recent large cycles. At this stage, such a persistence forecast is the best method of forecasting this cycle."

The next cycle may not be similar, but if the overall planetary model is valid, the positions of the major planets in a dynamo effect could offer indications for a more refined forecast. For example, with a mechanical generator an electrical engineer could easily calculate when the generator would commence charging and when charging would not take place. The problem is somewhat more complex in the planetary model because of the larger number of "magnets", ie. planets. In this, the faster moving planets would frequently be creating short-term charges but the full power would not be effective unless the most powerful "magnets" or major planets were in a suitable position of 90-degrees apart. With adequate knowledge, this also appears calculable. If the 60 and 120-degree alignments were an extension of the 90-degree principle, the major planets would continue to be effective as they open to a 120-degree position. As this is still within the dynamo principle, there could be an extended delay in the fall to the minimum and 2006 may well be an accurate prediction. Checks with short-term peaks showed that where there was a peak with no 90-degree alignments that a 60-degree or 120-degree alignment was often in that time-period. In this, they appeared to be equally in a position that was moving towards a 90-degree alignment as well as moving away from it, but the 90-degree alignment seems to be the key. The problem is that the more alignments we use the more there will be alignments to fill the gaps until it appears as only a coincidence. Apart from laboratory experiments, observation and analysis should help to clarify these points, and determine the process. This is essential for accurate predictions, because if predictions are from the assumption of a similarity to other large cycles there could be variations if a cycle is not similar, ie. it is "irregular". With a better understanding of how planetary radiation waves and magnetic fields interact to create cycles of energy, it could be possible to foresee such irregularities.

Future Alignments and a Unique Coincidence

As a means of offering a visual example, the diagram of alignments in Figure 40 is for the heliocentric positions of the major planets on the first day of every year from 2000 to 2007. This includes the date of 2004 for the minimum by the planetary model and the NASA/IPS date of 2006 based on expected similar cycles. The positions help to indicate when the sunspot count might increase in the ideal model of Jupiter and Saturn at 60 degrees apart for a new cycle. In addition, the diagrams indicate when a stronger effect should hypothetically occur from the 90-degree alignment and then to the further 120-degree position. Overall, the planetary model suggests a sharper drop to minimum than the official predictions do but if we are indeed in a period of irregularity, all the models could be inaccurate.

The tables and the diagrams show that the years 2003 and 2004 have major alignments that are well away from any 90-degree positions. In contrast, the year 2006 shows Jupiter making near 90-degree alignments to Saturn and to Neptune. Hypothetically, this would be a period of increasing sunspot activity, although similar situations have occurred before. On the face of it, the planetary model suggests a sharp drop to the minimum in 2004 and a possible steep rise to the following maximum, resulting in a shorter than normal period. If the minimum does occur in 2006 it would indicate that the planetary model is faulty or at least not accurate enough. Either way there would be an extra problem of methodology, but any anomaly may be because the longer Uranus/Neptune cycle had ended in an exact conjunction in the previous cycle. These outer two planets were exactly in line on April 20 1993, during the decline from maximum to minimum. There are other examples, of Uranus and Neptune closely in line that appear to disrupt the "normal patterns". These are with seismic effects. For example, in December 1811 and January and February 1812 there were three large earthquakes in New Madrid, Southern Missouri, that were felt as far away as Louisville, 300 kilometres away. Jupiter and Saturn did not fit the average prominent alignments at that time and it seems that a close alignment of Uranus and Neptune affects the energy rhythms. This outer alignment does not occur very often, and the events in the recent period of 1993 are therefore more likely to offer better detail for analysis of any effects.

As a further interest in the 1993 period, a graph for 1992-1994 in the IPS report (1995) showed the difference in sunspots in the northern and southern hemispheres[3]. The author stated that the reason why this sometimes occurs is unknown. The statistical comparisons of other differences with consideration of the relative position of the operative planets may perhaps offer a clue on this. This suggestion is of course speculative but it may relate to a planetary alignment. As it was over a period of three years, it appears to suggest a slow effect rather than a short-term effect. The IPS graph shows the two lines of northern and southern sunspot appearances well apart on the first day of 1992 with a difference in sunspot numbers between 60 for the northern spots and 100 for the southern spots. For April/May of 1993, the graph lines cross and show equal sunspot numbers with a count of below 40. The lines then continue down with a difference of about 10 until April/May 1994 when the northern line curves upwards and crosses again. The heliocentric positions of the outer planets, in degrees, on April 20 1993 were as follows, Earth 209, Mars 153, Jupiter 190, Saturn 322, and Uranus conjunction to Neptune 289. Later, in April/May of 1994, the graph lines changed back. This was where the south line again crossed the north line. Before that, the line turned upwards in September/October 1993.

The first coincidental phenomenon was in April/May 1993 where Mars was approximately 1.5 degrees north in celestial latitude. In the following year, in April/May 1994, Mars was approximately 1.5 degrees

---

[3] Variation of Hemispheric Sunspot Number 1992-1994. *Is The Solar Cycle Declining?* Richard Thompson. (Copyright 1995 IPS Radio & Space Services)

south in celestial latitude. At the halfway point, of October 3 1993, Mars was 0 degrees celestial latitude. That was approximately when the south graph line started to curve upwards. At the beginning of 1992 Saturn, Uranus and Neptune were near to zero degrees latitude. In degrees and minutes, Jupiter was 1 04 N, Saturn 0 38 S, Uranus 0 23 S and Neptune was 0 45 N. Later, in October 1993 Jupiter was 1 16 N and Saturn was 1 25 S. As these two planets seem to be significant in the sunspot process, there may be an explanation for these changes and coincidences. The second strange coincidence was that in April/May 1993 Mars was opposite to Saturn and in April/May 1994 Mars was conjunction with Saturn. At the halfway point in October 1993, when Mars was crossing the celestial zero latitude Mars was 90-degrees to Saturn. This was when the north graph line curved up slightly towards the south line. How this could be important is another mystery, as we have to consider the connection in terms of physics. Yet, it seems to be a very remarkable coincidence. The report indicated that there are other times when this phenomenon occurs, and other sources explain that from 1955 to 1970 there were more spots in the northern hemisphere, but in the 1990 cycle there were more in the southern hemisphere. This may relate to the outer planets, and the way their inclined orbits cause the radiation fields to interact, especially as the sun is also slightly inclined on its axis. The details appear to suggest that there is a specific linear interaction, in both latitude and longitude. A better analysis may produce more clues on the significance. By anticipating the alignments and knowing the connection with the changes in longitude and latitude, it may even be possible to foresee the differences in north/south sunspot numbers.

Apart from that, there are other unanswered questions. The previous maximum was in 1989 and the minimum was in 1996 although Jupiter and Saturn were 90-degrees apart in November of 1995, even though this was leading up to the period of minimum. The daily sunspot numbers were then less than ten, except when the faster moving planets gave a temporary boost, and in November, when no other planets were helping the Jupiter/Saturn alignment the count was zero. The important point here is that these different conditions do not match the simplified ideal planetary model with only Jupiter and Saturn as key planets and it seems that consideration of other alignments is necessary. The indication appears to be that the longer cycle is perhaps the dominant one and Uranus and Neptune can therefore disturb the ideal rhythms of the basic Jupiter/Saturn planetary model. Although other combinations seem significant, the Jupiter/Saturn alignments appear to govern the changing of solar and sunspot polarity. As the other planets may add to the process, we have to consider a wider framework.

The Ninety Year Cycle

All astronomers know that the concept of a general eleven-year sunspot cycle has been prevalent ever since Heinrich Schwabe announced his findings of the eleven-year rhythm in the 19th century. His conclusions of an eleven-year cycle are understandable, because it is easier to observe a number of such cycles in a lifetime of observation. The longer cycle of ninety years and especially the double cycle of one hundred and eighty years would escape detection by personal observation. The accumulated records that we now have give a better indication and show that the longer rhythm relates to Uranus and Neptune on the same principle as that discussed for Jupiter and Saturn. Therefore, if we expand the approach and take the Uranus-Neptune alignments as significant we can see that the anomalies might relate to such alignments. Any analysis of sunspot effects, such as earthquakes, weather, or other effects would therefore have to be in the longer time framework, otherwise, a fragment is mistaken for the complete picture. In relation to these points Uranus and Neptune were opposite in May 1908, 90 degrees apart in April 1955, conjunction in April 1993 and they will be 90 degrees apart in August 2040. If the consideration of effects is in this wider framework, the apparent variations are perhaps more understandable. The problem is complex because the indication is that all the major planets can act with each other as "dynamo magnets". In principle, there are the two pairs of magnets (planets) that statistically fit the model. Because of this, and the recent major conjunction of Uranus and Neptune, we are now apparently entering a phase of approximately sixteen eleven-year cycles in the greater one hundred and eighty-year cycle. There will of course be the halfway point in this major cycle where these two outer planets are opposite each other. If the model is correct, expert mathematical analysis could no doubt refine the timing and decide from that when key changes will occur.

The real problem seems to be in determining the type of radiation involved and deciding how it operates in the dynamo principle. There seem to be three basic principles that combine to make an integrated system. One is that of an electromagnet where an incoming current magnetises the core, and north and south depend on the direction of the winding. Another is with a generator where the armature sweeps through the field created by two magnets, and the north and south polarity relates to the order of the two magnets. The third, for which there is no clear model, appears to be some type of moving radiation system whereby the planets are sending out energy as well as acting as magnets. The big question is, how do they fit together whereby there is more than one system operating within the longer 90/180 years cycle?

Earthquakes and Related Effects

The map in Figure 11, showing major seismic activity for 1999, is not strictly typical of the spread of all earthquakes, as a wider range of magnitude shows a more extensive spread of earthquakes. The list for those events, in the Appendix, is included for the use of students and researchers who may wish to check any details. In comparison of magnitudes, there were approximately one hundred earthquakes of above magnitude 6, world wide, in that year but there were more than one hundred above magnitude 4.9 in just one month. That is to say, world-wide there were over one thousand above magnitude 4.9 during the year. This means there would proportionately be more seismic activity in any affected area.

The small circles show areas where three plates meet. However, there are not always earthquakes in all such areas. This shows that an association of seismic activity with biological disturbance is not at all clear. Effects might be from ionisation caused by horizontal compression, without seismic activity, or even from an associated disturbance of the ionosphere.

Also, variations from higher tides seem to occur when the moon is in line with a planet, especially Venus, Mars, Jupiter and Saturn, but usually when they are at their nearest to the earth. Such alignments appear to match graph peaks of seismic activity as in Figure 37. On this hypothesis, graph peaks of recorded tide-levels would match daily alignments of the moon to planets. Confirmation of this could perhaps help to clarify how far external traction triggers seismic activity as opposed to the internal effects. With a higher tide, the horizontal compression from the tides would probably be greater. A simple check might be with seismogram records where the daily oscillations in seismic activity could match against daily conjunctions and daily tide levels. Hypothetically, they should all be similar. In any case, scientists know that ocean tide variations noticeably match seismic activity in many areas. Despite that, these have not considered the addition of planetary traction as an extra tide raising force. The point is that although general fluctuations broadly match the sunspot activity the individual seismic peaks appear to relate to extra tidal effects where planets add to the lunar traction. In this sense, there could also be a type of Spring Tide effect where the moon or sun is opposite a major planet or group of planets, which are at their nearest point to the earth. As an example, the moon was opposite to Jupiter and Saturn, which were at their nearest to the earth, at the time of the Turkey earthquake in

August of 1999. There are other similar patterns with Jupiter and Saturn at their nearest, such as in November 2000. In geocentric alignments, the sun was opposite Saturn on November 19 and opposite Jupiter on November 29. When an outer planet is on the opposite side of the earth to the sun, the earth is on the planet side of the sun and therefore nearer to the planet. The gravitational force will consequently be at its maximum, but the greatest effect during such a period would be when the moon is in line with the planet. The moon was conjunction to Saturn on November 12 and conjunction to Jupiter on November 13. Effects are apparently not instantaneous, but on November 16, there were 30 disturbances above magnitude 4.9. There were many floods world wide in that period, and observers in tide affected areas might find it of interest to check actual recorded tide levels with earthquakes. In that period, the Full Moon Spring Tide was on November 11, the third quarter, Neap Tide, was on the 18th and the New Moon Spring Tide was on November 30. The most effective point appears to be the conjunction of the moon to Saturn and Jupiter, as allowing for any delay, there was no high peak of sunspot numbers that appears to match it. This seems to suggest that the lunar and Jupiter and Saturn conjunctions are a contributory trigger for the jump in seismic activity on the 16th. Comparison with ocean tides levels near this date may help to clarify the questions, as almost all of the thirty seismic disturbances were in the area of the Solomon Islands.

Other interesting tidal patterns occur with many seismic events but they are often so intricate that concise explanations are difficult. Very often, there is not a sharp distinction between a tidal effect and other contributory causes. For example, another related effect might be the differences caused by atmospheric pressure. Scientists know that high pressure can help to reduce the tide levels, and from this we can see that low pressure would help the tide to be higher. Therefore, if there is high pressure in one area, depressing the ocean and low pressure in another area, there could be a higher tide than normal in the low-pressure area.

This would appear to suggest that if other factors were present a period of low pressure would add to the risk of seismic activity by increasing the mass of water in the critical area and adding to the horizontal compression.

Another seismic phenomenon is the occurrence of spates of minor earthquakes, which are usually below magnitude 5. For example in 1965-1967 there was continual activity with many thousands of such tremors. The peak was on April 17 1966 with thousands on that day. This seems to match a series of ninety-degree alignments with the earth at the focus and at the peak Jupiter was almost at ninety-degrees to Saturn. Whichever way we examine the problem, there appears to be a connection with the planets in terms of gravitational forces and electromagnetic effects.

Further Volcanic Activity

There were further volcanic eruptions in the year 2000. Each eruption is listed below as a separate event, as with earthquakes.

January	30th; Nyamuragiri Congo.
February	12th; Kamachatka, Russia, 14th; Island of Reunion, 23th; Kilauea, Hawaii, 26th; Hekla, Iceland.
March	1st; Pacaya, Guatemala, 11th; Mayon, Philippines, 15th; Bezmianny, Russia.
April	16th; Pichincha, Ecuador, 17th; Usu, Japan.
May	15th; Shishaldin, Alaska, 20th; Ecuador, 12th-19th; Monserrat, West Indies, 13th; Nicaragua, 14th; Kavachi, Solomon Islands, 17th; White Island, New Zealand, 23th; Popocatepetl, Mexico, 29th; Cameroon.
June	1st; Etna, Sicily, June 7th Cameroon-Solomon Islands, 15th Etna-Sicily, 23th Island of Reunion

July	12 th Ecuador, 12 th Mount Oyama-Japan, 16 th Copahue-Argentina, 28 th Kamchatka, 30 th Mayon-Philippines, Continuing Semeru-Java.
August	10 th Popocatepetl-Mexico, 6 th Kamchatka, 16 th Honsu-Japan, 17 th Kaba-Sumatra, 23 rd Arenal-Costa-Rica, 28 th Etna-Sicily, 31st Oyama-Japan.
September	6 th Papua New Guinea, 8 th -15 th Montseratt-West Indies, 21 Guatemala, 21 Popocatepetl-Mexico, 28 th Hokkaido-Japan.
October	9 th Sakura-Jima Japan, 12 th Island of Reunion, 29 th Popocatepetl-Mexico.
November	8 th Komagatake-Japan, 10 th Colima-Mexico, 11 th Popacatapetl-Mexico, 11th Karangetang-Japan, 15 th - 21st Merapi-Java.
December	9 th Fuego Guatemala, 19 th Popocatepetl Mexico.

The total number of volcanic eruptions in 2000 is 48. The year of Sunspot Minimum, in 1996, had 10, 1997 had 15, and 1998 had 16. The month of May appears to be a period of increased activity, with a total of 8. This fits the hypothesis, that ninety-degree alignments and close groupings affect terrestrial activity.

The table below shows the steady increase in general volcanic activity as the sunspot cycle progressed from Minimum to Maximum. The highest peak in the sunspot activity was in July 2000, and the highest peak in general volcanic activity was in August 2000, which matches the general delay from sunspot to terrestrial activity.

MONTHLY TOTALS OF RECENT VOLCANIC ERUPTIONS

Year	Jan	Feb	Mar	Apr	May	June	July	Aug	Sept	Oct	Nov	Dec
1999	2	3	0	1	0	3	3	2	2	1	4	1
2000	1	4	3	2	8	4	4	7	5	3	5	2

(Info Internet Nodak/Lycos)

YEARLY TOTALS OF VOLCANIC ERUPTIONS

Year	1995	1996	1997	1998	1999	2000
Total	5	10	15	16	22	48

One particular example of weather effects from volcanic activity has emerged from research into weather conditions in the sixth century. Worldwide cooperative investigation narrowed down the likely cause to be an earlier explosion of Krakatoa. As pointed out earlier the Krakatoa area is vulnerable to excessive horizontal compression from tidal pressures. The date arrived at, from tree ring analysis and carbon dating of samples, was approximately 536 AD. The weather was then apparently abnormally cold, with the sun obscured for years. In terms of the planetary model, the alignments and grouping of the major planets were critical, and it could well have been a period of strong seismic activity, but in the absence of exact times, analysis is difficult.

The increase in recent volcanic activity, in the approach to sunspot maximum, appears to confirm that there is an associated effect. This does not show with the study of only one volcano but is noticeable with the earth as a whole. The same problem exists with earthquakes and it is obvious that the study of only one seismic area would not reveal the external connection. We could also apply the sunspot effect to weather on the earth. If the earth is hotter at sunspot maximum because of extra activity in the sun and because of extra activity in the earth, the extra warming could affect the earth as a whole. The oceans and vegetation would presumably give off more evaporation, which would return as extra rain in different areas although some areas would apparently experience drought. Even so, knowledge of the overall process does not indicate which area will be the affected area and the final evaluation will always be from specialists in the different disciplines and on the spot observations. Nevertheless, it might be possible to foresee changes in these different activities with the aid of astronomical knowledge. There may be enough records from the past two hundred years to show if changes in temperature and rainfall match the long- term alignments. We might then find that there are natural cycles of change that are normal and our own contributions by chemical emissions are an extra factor. In any case, there was probably some effect from internal heart and volcanic ash and smoke in the atmosphere from the increased volcanic activity leading up to the sunspot maximum of 2000. Some of the global warming may therefore be an extension of the planetary effects and the present level could drop naturally. As there are other such periods, comparisons should be possible.

A paper entitled *The Influence of Planetary Bodies and Sunspots on Volcanic Activity (2002)*, gives more examples of the sunspot connection. (Paper available via Darch Literary Service Website: <u>darchliterary.iinet.net.au</u>).

Conclusion

The further phenomena that occurred during the year 2000 appear to have confirmed the general principles. The general hypothesis embraced two main principles, which were that the planets affect both the sun and the earth by a similar type of radiation and that there is a unique gravitational effect. Observation and investigation indicated the probable nature of the radiation and pointed to a very specific gravitational effect. The statistical data eventually indicated that there were definitely significant common elements with the sun and the earth. Even so, the task of establishing a more exact determination is very complex and the technical process will obviously need a team of experts to define the more ordered parameters for investigation. The statistical analysis only indicates the likely range of effects, but the processes are capable of simulation in a laboratory and scientists could probably refine the hypothesis and define the processes. A suitable dynamo model with multiple magnetic fields on different cyclical periods would certainly clarify the irregularities caused by the slower outer planets.

Observations by astronomers and geophysicists will no doubt clarify the practical usefulness of the model presented. The further development of the hypothesis definitely indicated that there is some type of electromagnetic interaction between the planets and the sun and demonstrably indicates that there is a recognisable "dynamo effect" that relates to planetary alignments. With the gravitational effect, three specific effects emerged. One is the application of the Indirect Traction and another is the combination of traction from external bodies. The third effect is the relative proximity of some planets at times of seismic events. There is a possibility that the relative direction of the planet to the earth creates an extra potential gravitational force affecting the earth, as with a Doppler effect.

The sample range studied, from the past 100 to 150 years, was well within access of accurate records and from these a more expert mathematical analysis could probably clarify the intricate variables. In addition, experienced observers will be able to check the principles against their practical field knowledge because overall they are within the formal framework of conventional physics. With some refinement, this could help in foreseeing solar and seismic activity. Although statistical analysis does not always lead to conclusions with certainty, one fact remains. That is, when the sunspots increase earthquakes will increase.

The key factor appears to be the planetary alignments, but if the generator principle is correct, there will always be some effect from the planets. The 90-degree alignment is probably the optimum position, but a mechanical generator would always produce some charge with closer or wider angular positions of the magnets. The verification of the main hypothesis rests on the clarification of how this process really operates. The secondary hypothesis, that indirect traction is the trigger for earthquakes, is only a process of applied mathematics. Both hypotheses are therefore within the range of formal physics, and can be confirmed or rejected in that framework.

Diagrams

Chapter 13

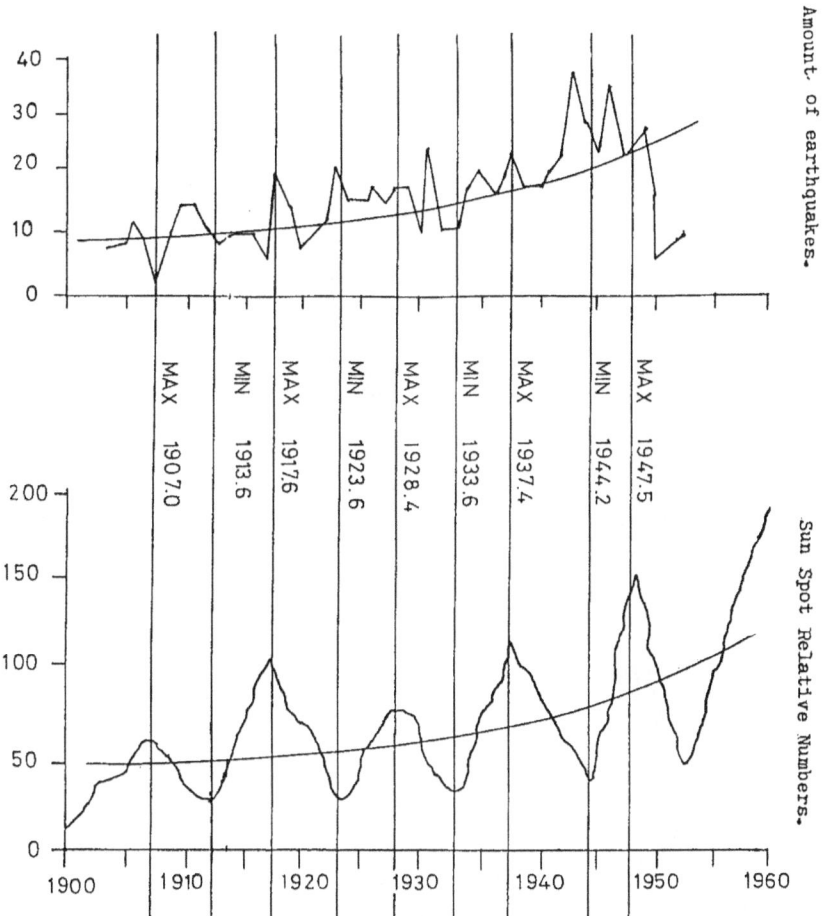

Major earthquakes and sunspot activity from 1900 to 1950
Showing similar smoothed curves.

Figure 1

Source–Earthquakes: Gutenberg, B. & Richter, C.F. *Seismicity of the Earth &*
Associated
Phenomena, Princeton University Press. (1954)
Sunspots: *Solar Data*, National Bureau of Standards, Colorado, USA (1964)

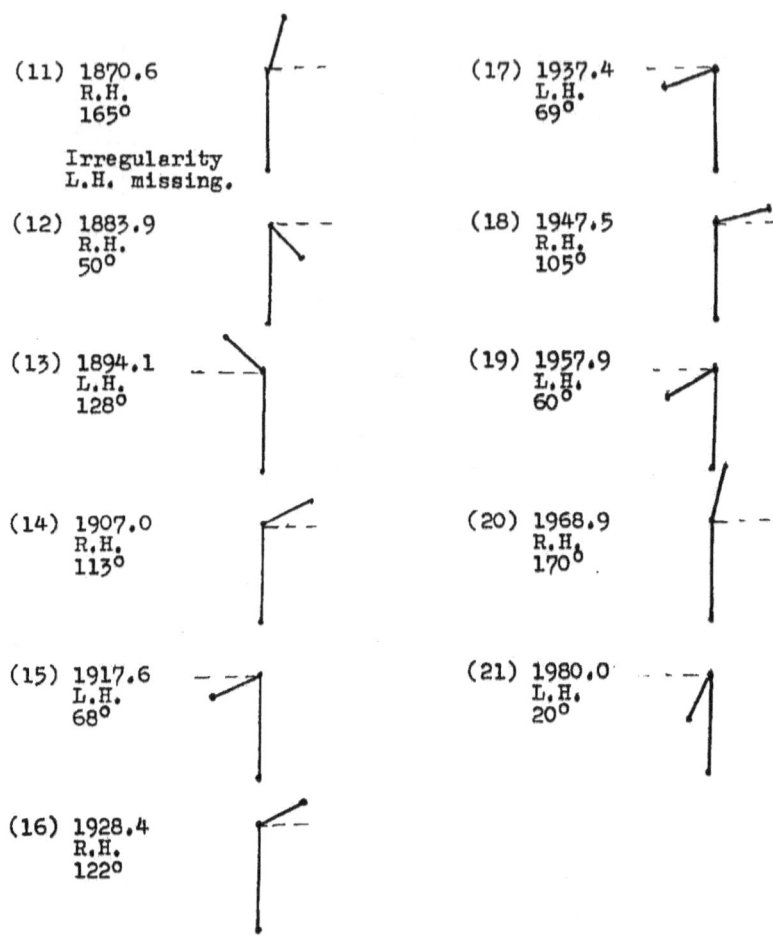

(11) 1870.6
R.H.
165°

Irregularity
L.H. missing.

(12) 1883.9
R.H.
50°

(13) 1894.1
L.H.
128°

(14) 1907.0
R.H.
113°

(15) 1917.6
L.H.
68°

(16) 1928.4
R.H.
122°

(17) 1937.4
L.H.
69°

(18) 1947.5
R.H.
105°

(19) 1957.9
L.H.
60°

(20) 1968.9
R.H.
170°

(21) 1980.0
L.H.
20°

Relative positions of Saturn and Jupiter at Sunspot Maximum

Saturn is the long arm. The sun is at the focus. The planets orbit anticlockwise. After
Cycle No 11 the Saturn-Jupiter alignment was in 1878.9. The Sunspot Maximum for
Cycle No 21 was in December 1979. It gave a smoothed relative number of 164.5. The
Saturn-Jupiter alignment after that was in April 1981.

Figure 2a

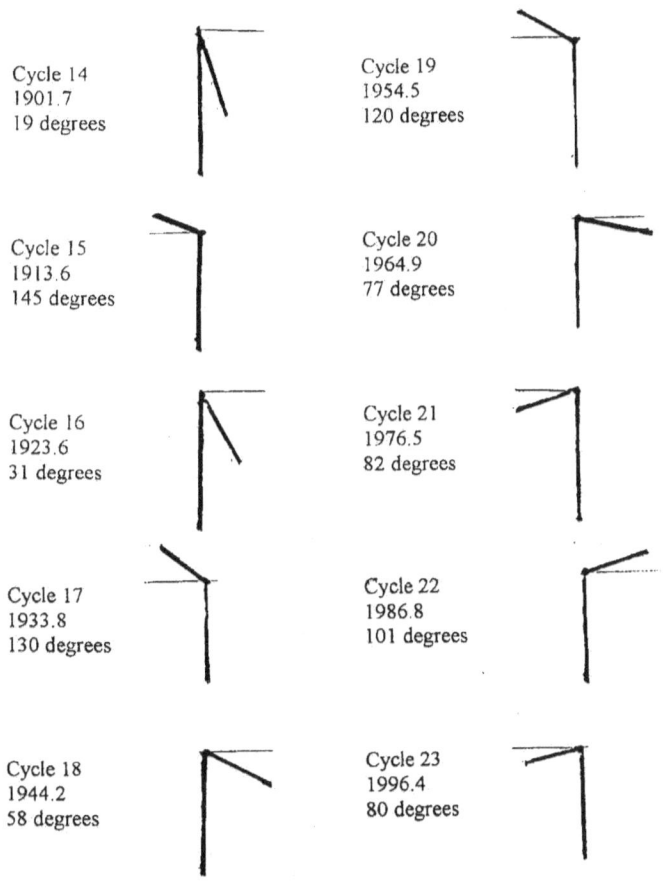

Cycle 14
1901.7
19 degrees

Cycle 15
1913.6
145 degrees

Cycle 16
1923.6
31 degrees

Cycle 17
1933.8
130 degrees

Cycle 18
1944.2
58 degrees

Cycle 19
1954.5
120 degrees

Cycle 20
1964.9
77 degrees

Cycle 21
1976.5
82 degrees

Cycle 22
1986.8
101 degrees

Cycle 23
1996.4
80 degrees

Relative position of Jupiter and Saturn at Sunspot Minimum

Saturn is the long arm. The planets orbit anti-clockwise.

Figure 2b

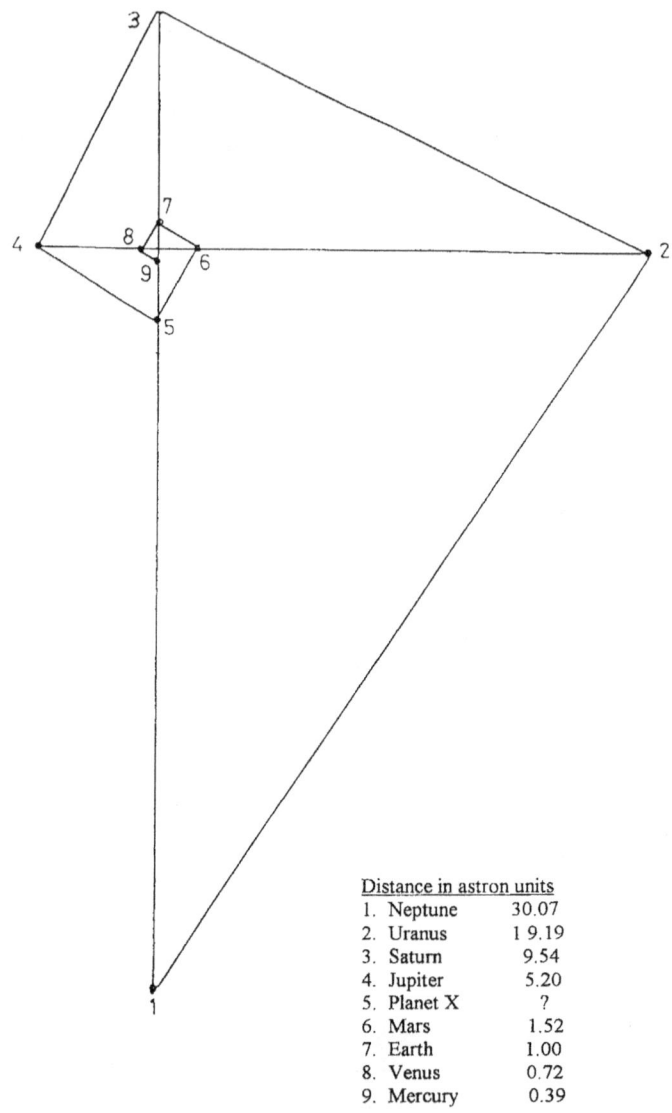

Distance in astron units	
1. Neptune	30.07
2. Uranus	1 9.19
3. Saturn	9.54
4. Jupiter	5.20
5. Planet X	?
6. Mars	1.52
7. Earth	1.00
8. Venus	0.72
9. Mercury	0.39

Ideal positions

Figure 3

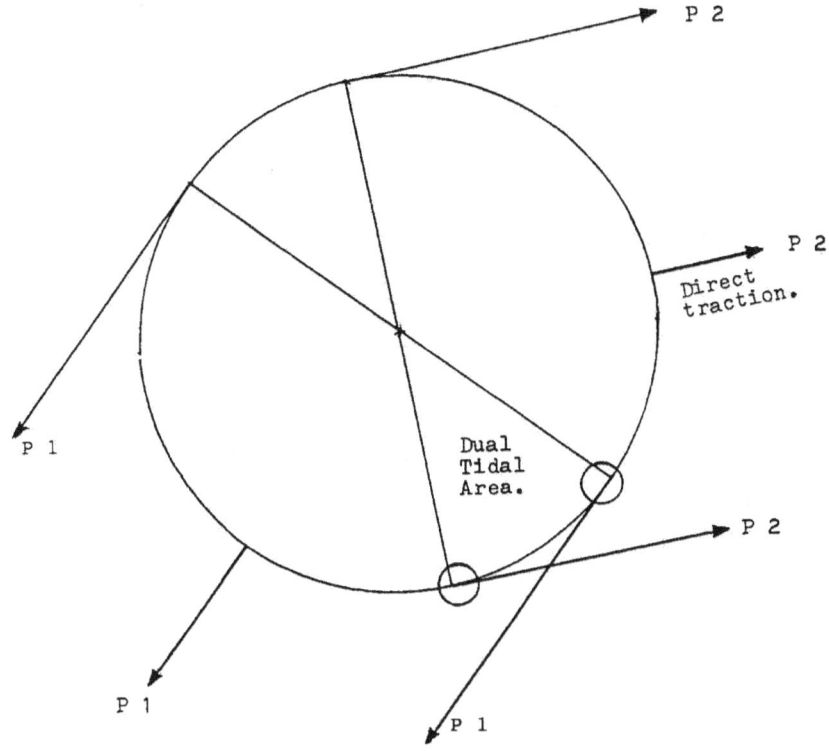

<u>Example of overlapping traction fields</u>

The flares generally occur at the edge of the
overlapping fields, as indicated by the small circles.

Figure 4

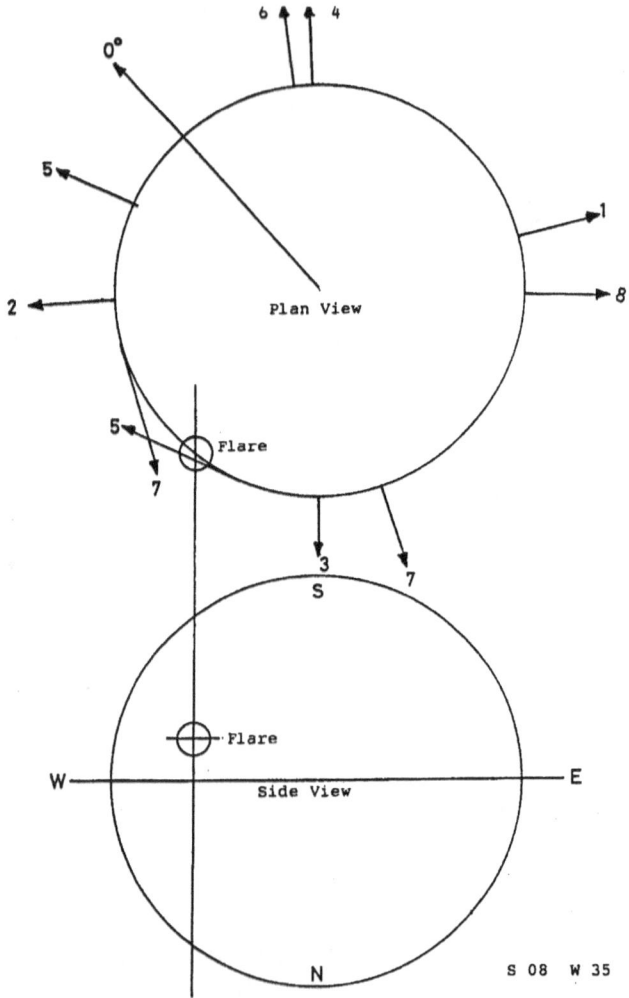

Solar flare. February 8th 1964

1 Mercury, 2 Venus, 3 Earth, 4 Mars,
5 Jupiter, 6 Saturn, 7 Uranus, 8 Neptune

Figure 5a

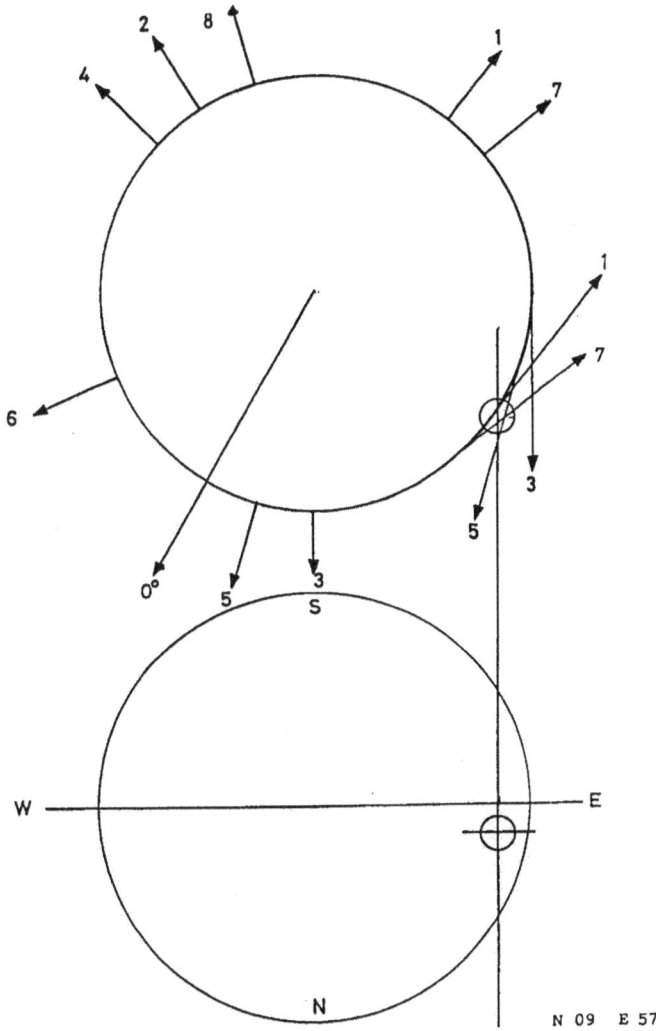

<u>Solar flare. October 22nd 1963</u>

Figure 5b

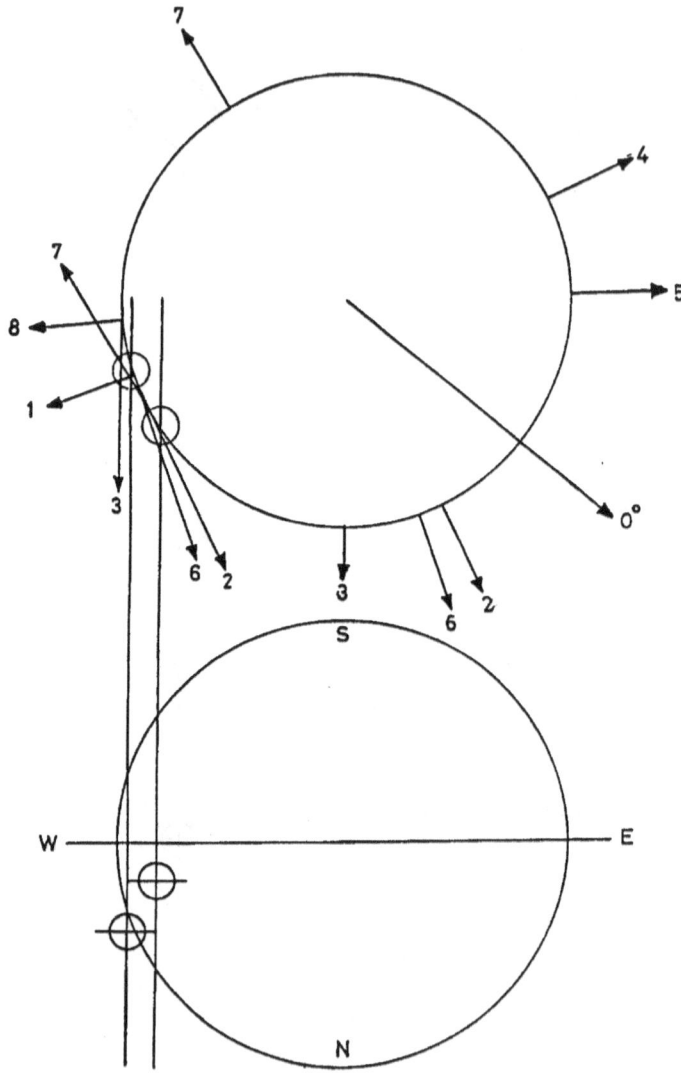

A double flare. August 1st 1964

Figure 5c

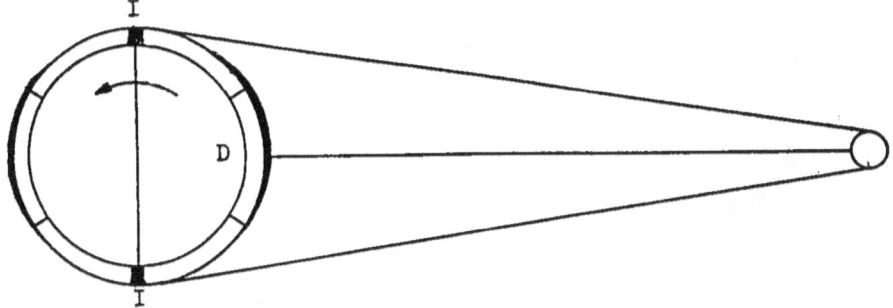

<u>Areas of seismic stress—points of direct and indirect traction</u>

E = External Body. D = Direct Traction. I-E = Indirect Traction.
I = Point of Stress. I-D = Total Traction Area.

The six sections represent the six main tectonic plates. When
a plate passes the point "I" it experiences stress as it enters
or leaves the traction field of the external body or bodies.

Figure 6

Note: One tectonic plate is only approximately one hundredth of the earth's total
mass. At the point of entry or of leaving the external gravitational field, the forces
acting on the plate do not have to compete with the earth's total gravitational pull, as
at that point the external forces bypass the contrary pull of the earth. In addition, the
plate easily moves sideways, as opposed to moving outwards, as with Direct Traction.
The relative mass of the moon is 0.0123 of the earth's mass and therefore has little
effect unless helped by the sun or a planet, especially Jupiter, Saturn and Venus at
their nearest to the earth. These are the critical points in time, which specialist computer analysis could calculate.

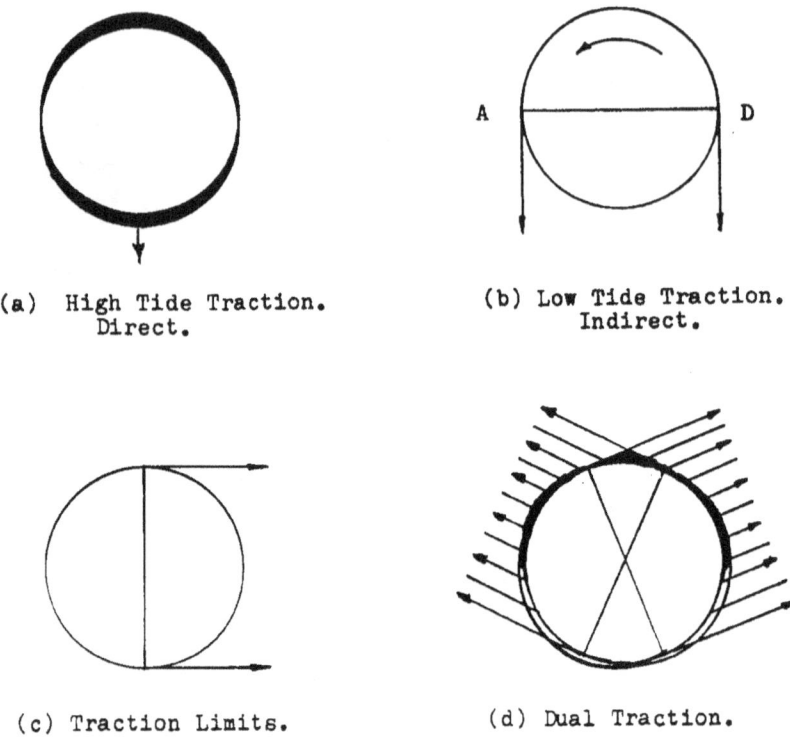

(a) High Tide Traction.
     Direct.

(b) Low Tide Traction.
     Indirect.

(c) Traction Limits.

(d) Dual Traction.

Principle of direct and indirect traction

(a) This shows the high ocean tide caused by direct traction and the complementary high tide on the opposite side of the earth.

(b) The points A and D show the low tide areas and indicate where the faults enter or leave the traction field. There would be acceleration at point A and deceleration at point D.

(c) This shows the limits of the traction field over one half of the earth's globe.

(d) This shows the contrary effect of traction from opposing directions. The darkened triangular area is the point of maximum stress.

Figure 7

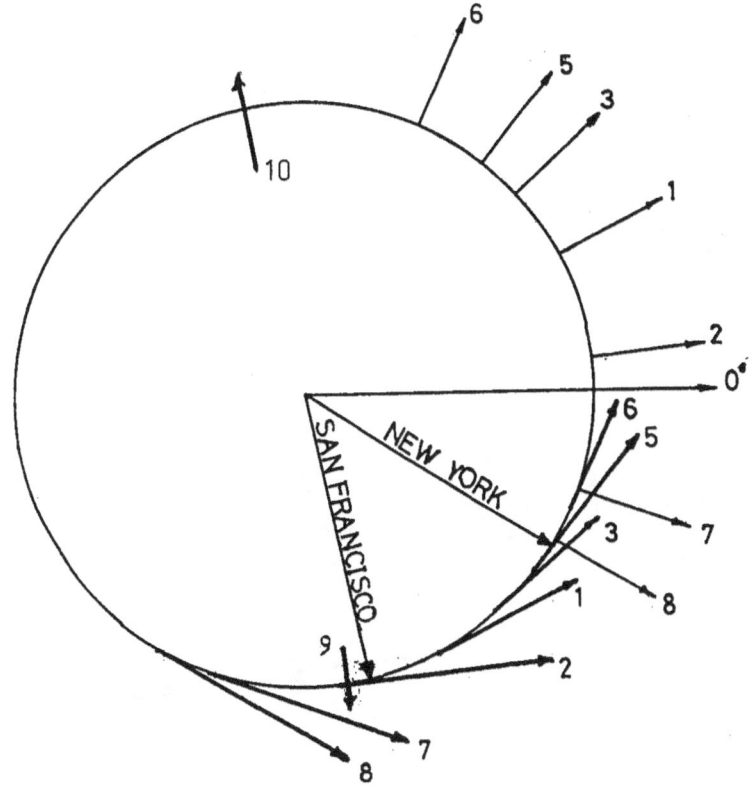

<u>San Francisco earthquake</u>

38 N. 123 W. 13:12 U T. April 18th 1906

Showing Indirect Traction applying to the area of
the seismic disturbance.

1 Sun, 2 Mercury, 3 Venus, 4 Earth, 5 Mars,
6 Jupiter, 7 Saturn, 8 Moon, 9 Uranus, 10 Neptune.

Figure 8

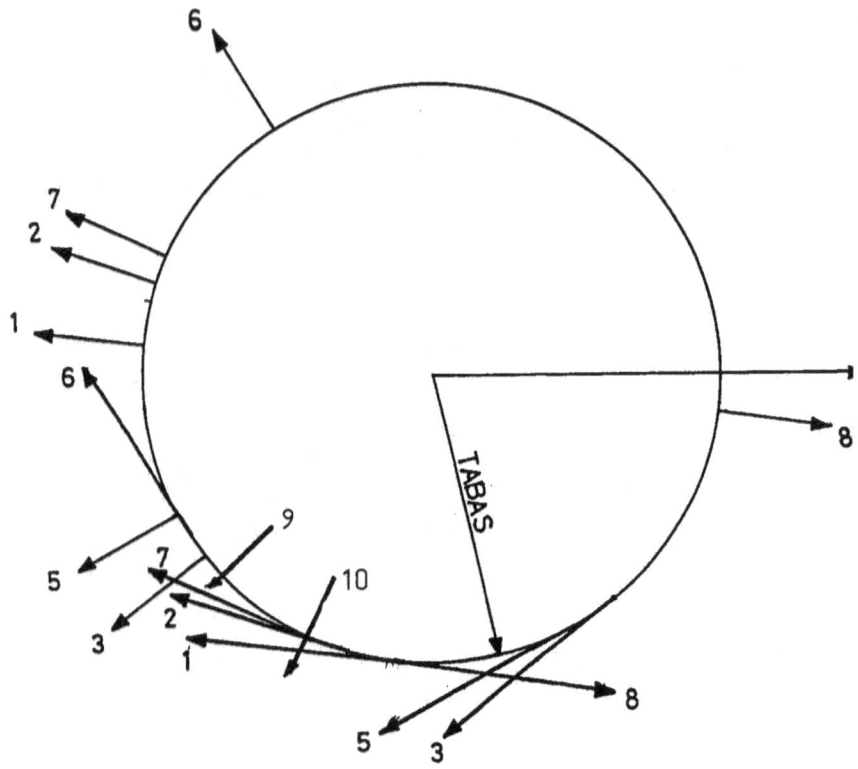

Tabas Iran earthquake

33 N. 57 E. 15:38 U T. September 16th 1978

Showing indirect traction applying to
the area of the seismic disturbance.

I Sun, 2 Mercury, 3 Venus, 4 Earth, 5 Mars,
6 Jupiter, 7 Saturn, 8 Moon, 9 Uranus, 10 Neptune.

Figure 9

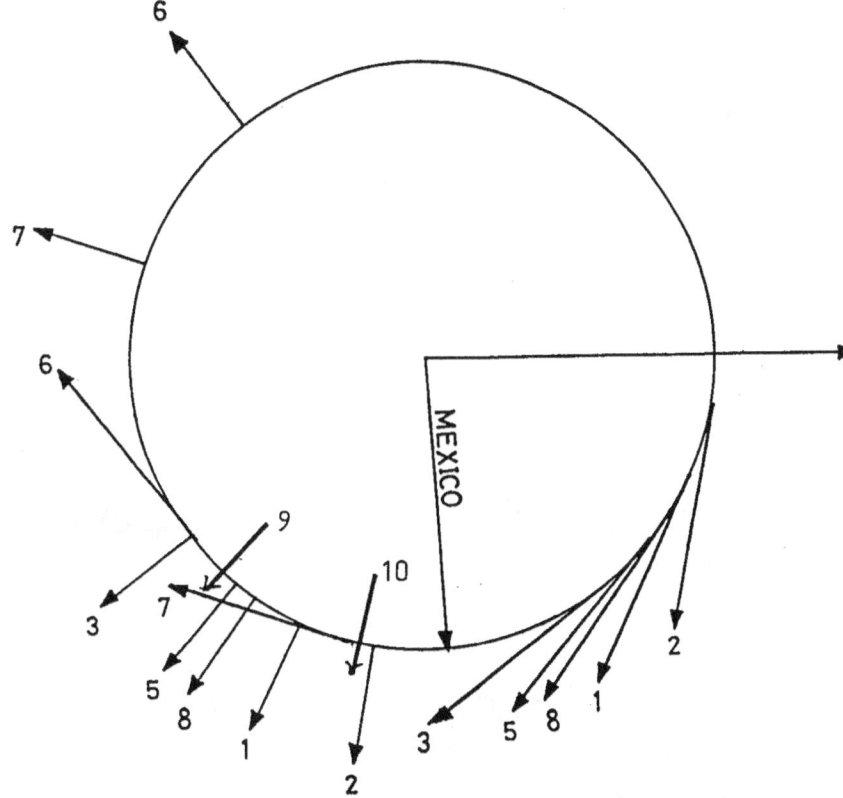

Mexico City earthquake

19 N. 99 W. 20:21 U T. November 29th 1978

Showing Indirect Traction applying to
area of the seismic disturbance.

I Sun, 2 Mercury, 3 Venus, 4 Earth, 5 Mars,
6 Jupiter, 7 Saturn, 8 Moon, 9 Uranus, 10 Neptune.

Figure 10

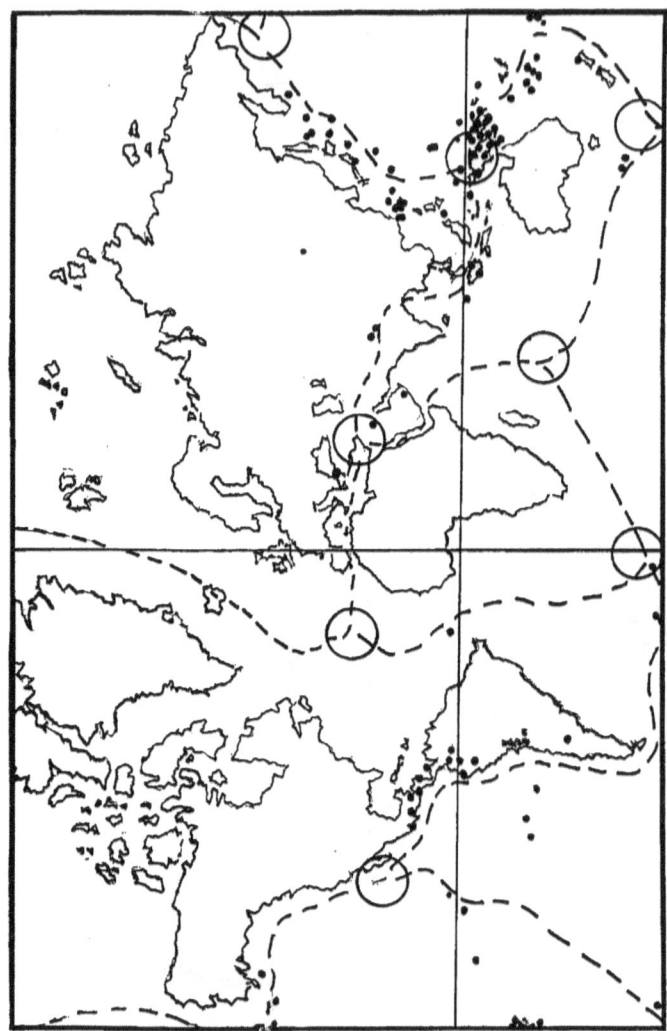

<u>Earthquakes of magnitude 6 and above for 1999–Total 102</u>

Dotted lines are fault lines; circles are main intersections where tectonic plates meet.

Figure 11

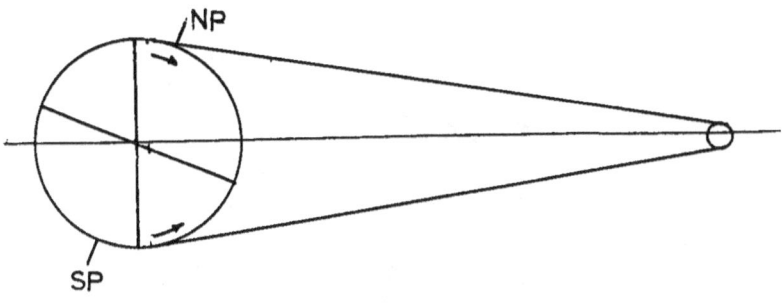

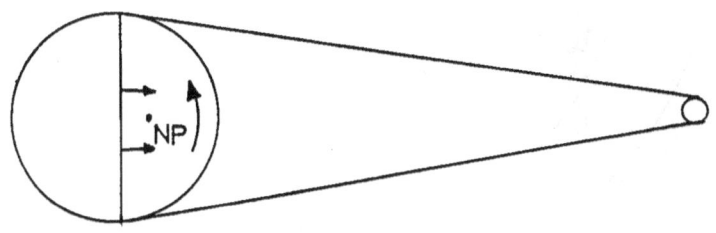

Indirect traction on the polar region

Showing relative action of external forces on
Glacier movements

Figure 12

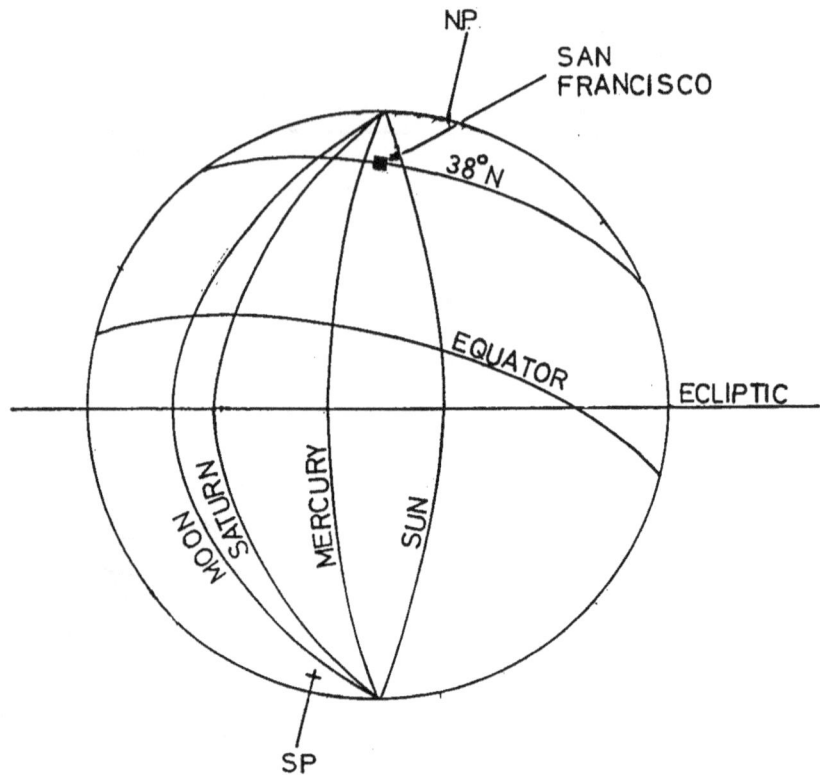

Latitude side view for San Francisco

Showing traction limits in relation to the
position of the fault

Figure 13a

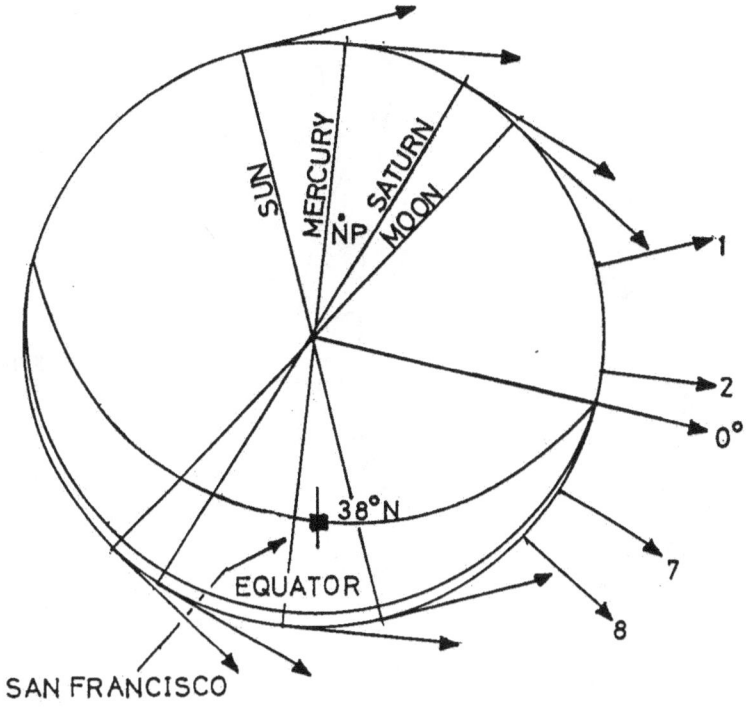

<u>Latitude plan view for San Francisco</u>

Showing traction limits in relation to the
position of the fault

Figure 13b

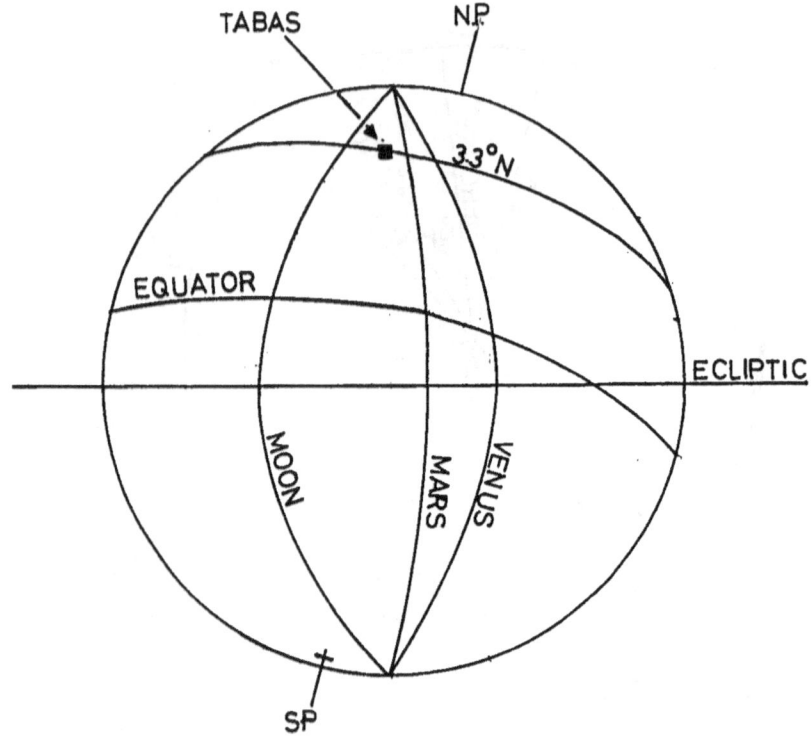

Latitude side view for Tabas

Showing traction limits in relation to the
position of the fault

Figure 13c

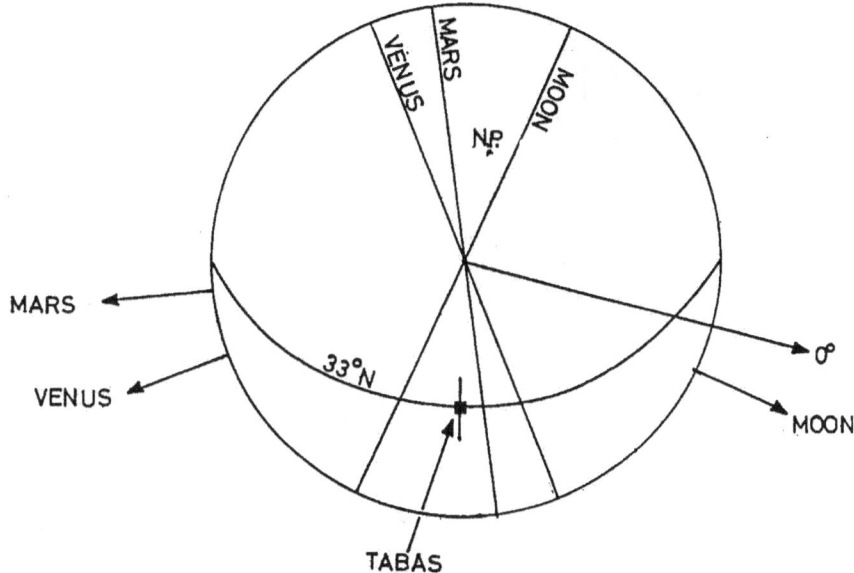

<u>Latitude plan view for Tabas</u>

Showing traction limits in relation to the
position of the fault

Figure 13d

(A)

(B)

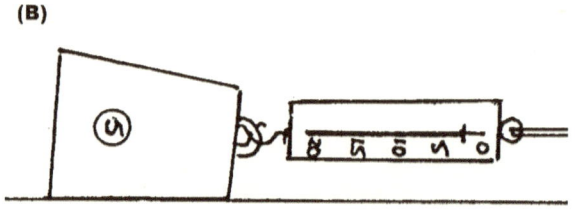

Example of gravity effect

Showing relation to the line of force

(A) Direct Traction: when pulling directly against the earth's gravity the scale shows 5 kilograms of the weight.

(B) Indirect Traction: when pulling along the earth's surface, at right angle to the earth's pull, the scale shows a noticeably lower reading. This varies according to the roughness of the surface and the consequent resistance.

Figure 14

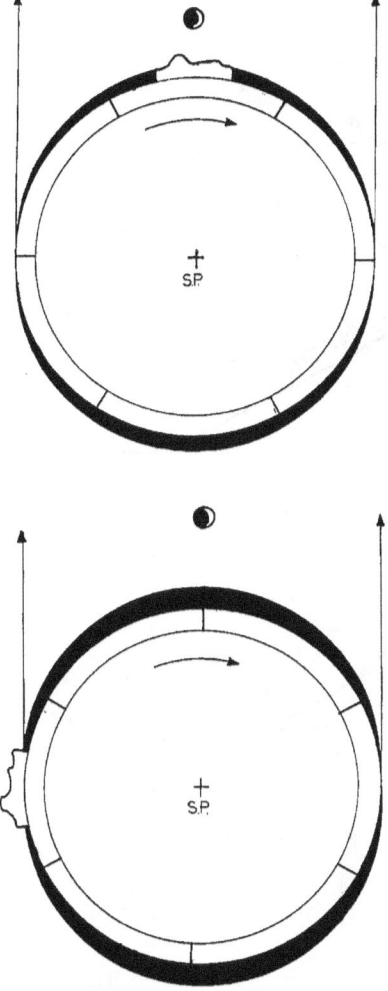

<u>Example of high and low tide related to a landmass</u>

Showing how tidal pressures create horizontal compression and
how less tidal pressure allows the land to move.

Figure 15

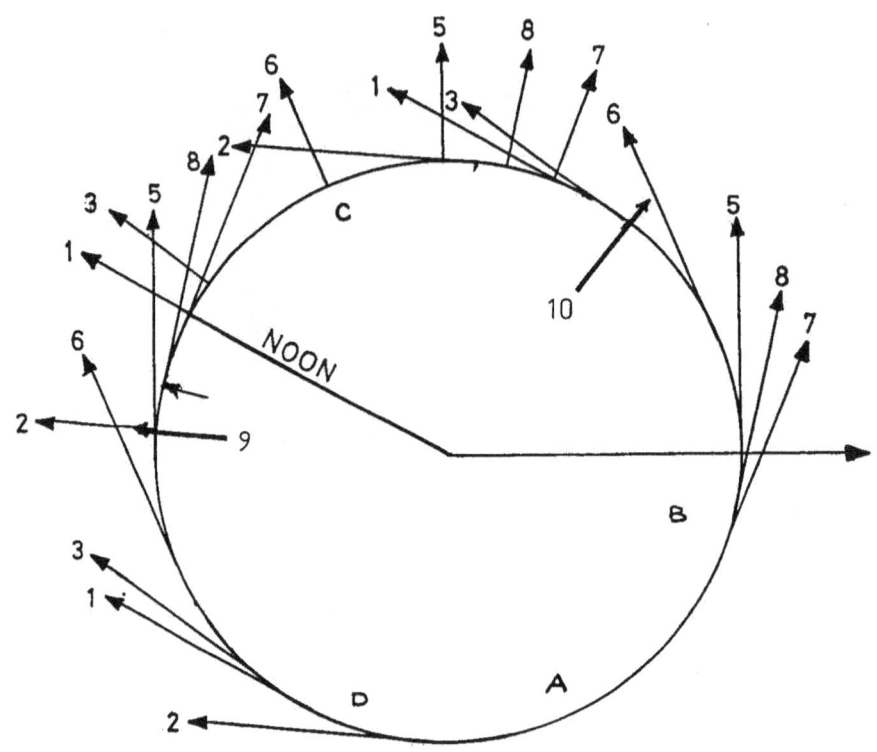

Explosion of Krakatoa volcano

1 p.m. August 26th 1883

The small arrow shows the position of Krakatoa
(The moon was just setting)
1 Sun. 2 Mercury. 3 Venus. (4 Earth). 5 Mars. 6 Jupiter. 7 Saturn. 8 Moon.
9 Uranus. 10 Neptune.

Figure 16a

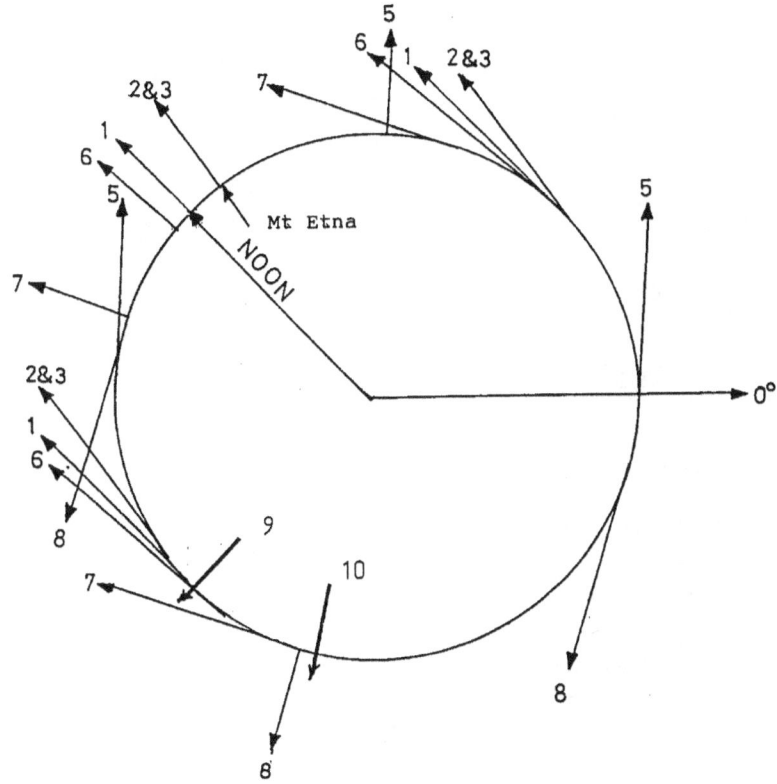

Explosion of Mt Etna volcano

11:30 a.m. August 5th 1979

The small arrow inside the circle indicates
the position of Mt Etna at 11:30 a.m.
The moon (No 8) was rising approximately
two hours later.

(Planetary code as before)

Figure 16b

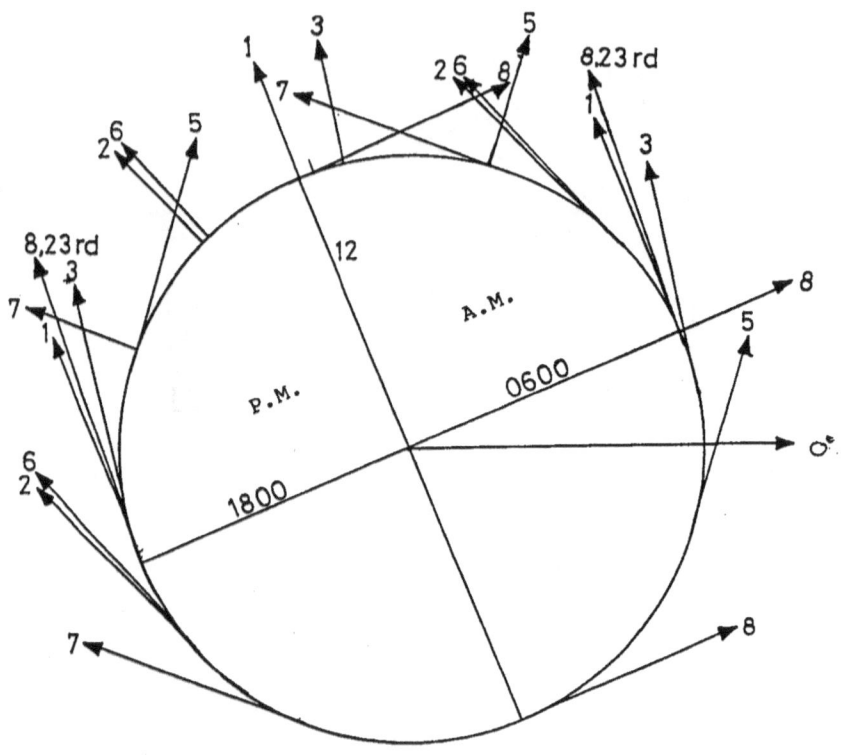

Earlier alignments for Mt Etna

July 16th 1979 a.m.
37.50 N 14.52 E

The rising and setting positions of the moon (No 8)
are also shown for the 23rd. It would be a New Moon Spring Tide.
On the 16th the moon was in the last quarter. It was a Neap Tide position.
In mid morning the affected area would experience cross traction
from the moon and Jupiter (No 6).

(Planetary code as before)

Figure 16c

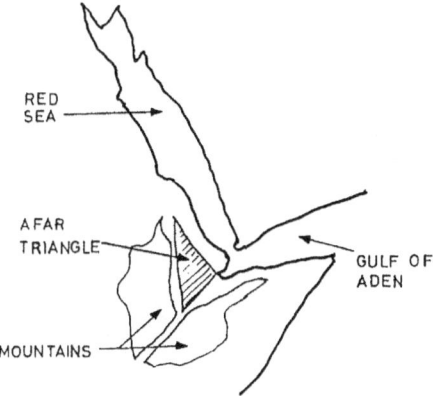

Tidal Pressures in the Gulf of Aden and Mt Etna

Showing position of mountains relative to the Afar Triangle and
relative position to tidal forces near the Straits of Messina.

Figure 17

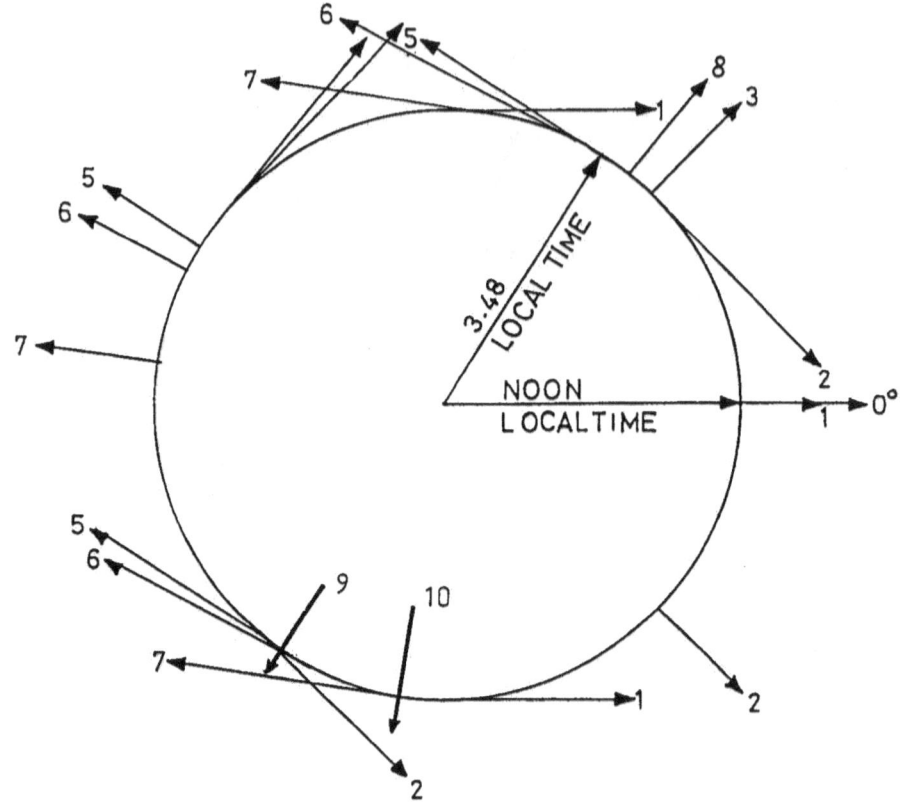

Mt St Helens volcano
Washington USA

Showing position at 3:48 p.m. March 20th 1980.
This was the time of the first tremor.
Mars (5), Jupiter (6) and Saturn (7) just rising.
The moon (8) was overhead,

Figure 18a

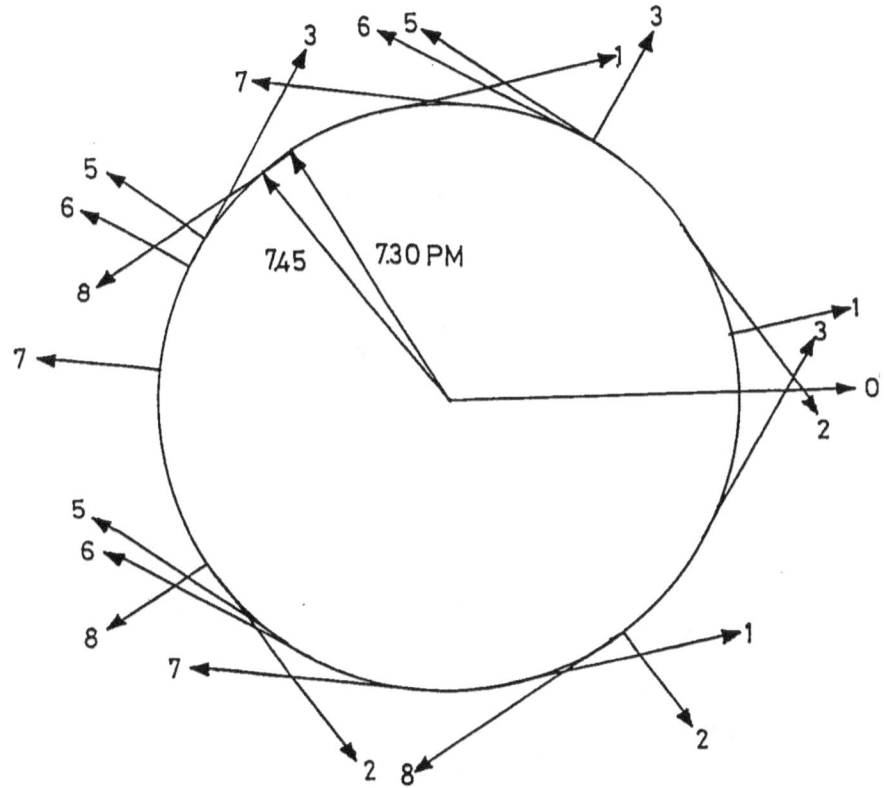

<u>Mt St Helens volcano</u>
Washington, U.S.A.

Showing position at 7:45 p.m. April 2nd 1980.
This was a period of harmonic tremors.
The moon (8) was just rising, and Venus (3) was near setting.
The sun (1) was setting at approximately 6:30 p.m. Local Time.

(Planetary code as before)

Figure 18b

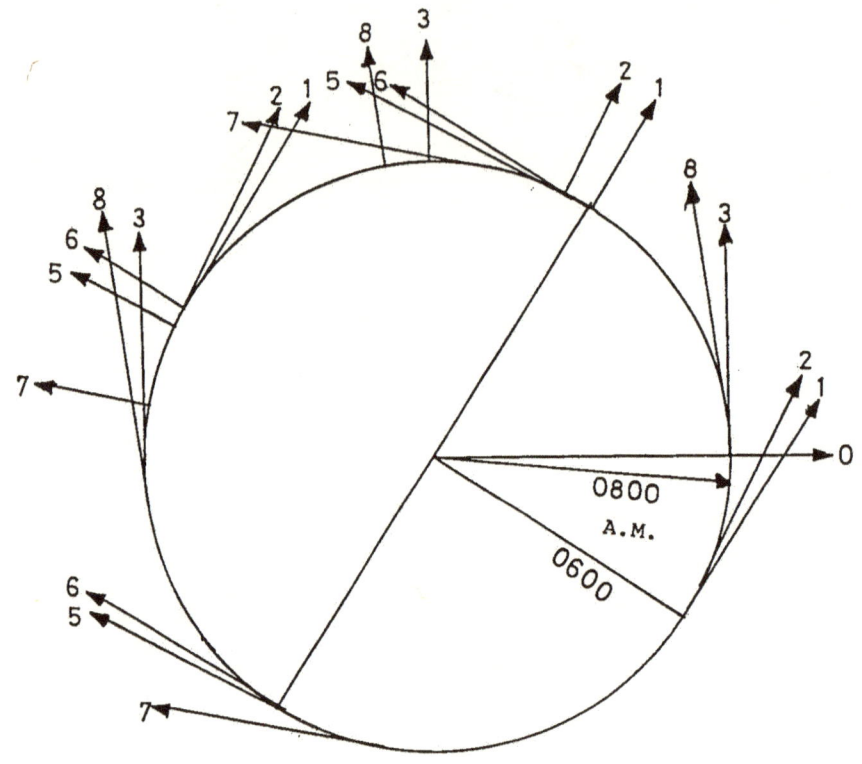

Mt St Helens volcano
Washington, USA

Showing position at 8:00 a.m. May 18th 1980
at time of explosions.

The Sun (1) and Mercury (2) had recently risen. The
Moon (8) and Venus (3) were near rising.

Figure 18c

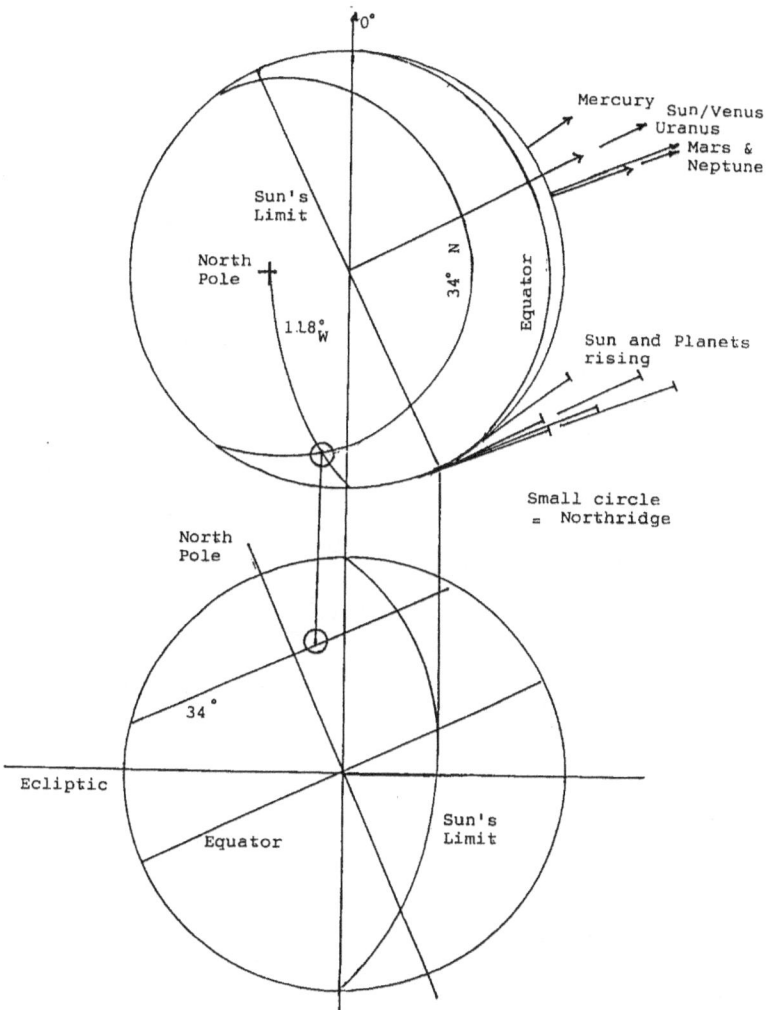

Northridge earthquake

January 17th 1994
Simplified plan and side view.

Figure 19

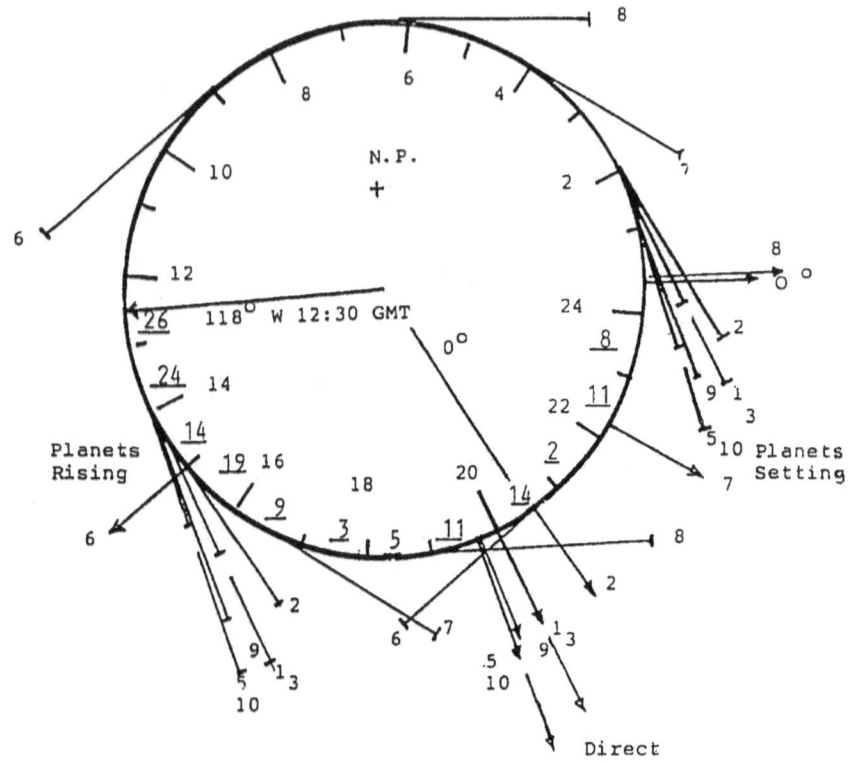

<u>Northridge LA earthquake</u>

34N 118W. 12:30 GMT. January 17th 1994
Showing Indirect Traction to the earth.
The inner arrow shows the position at 12:3 0 GMT.
The arrow moves anti-clockwise with the earth.

1 Sun, 2 Mercury, 3 Venus, 4 Earth, 5 Mars,
6 Jupiter, 7 Saturn, 8 Moon, 9 Uranus, 10 Neptune.

Figure 20

DAILY EARTHQUAKES FOR AUGUST 1999

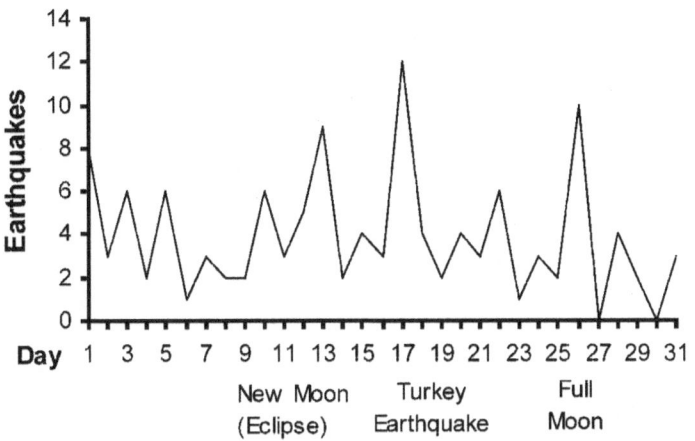

New Moon   Turkey   Full
(Eclipse)   Earthquake   Moon

DAILY EARTHQUAKES FOR SEPTEMBER 1999

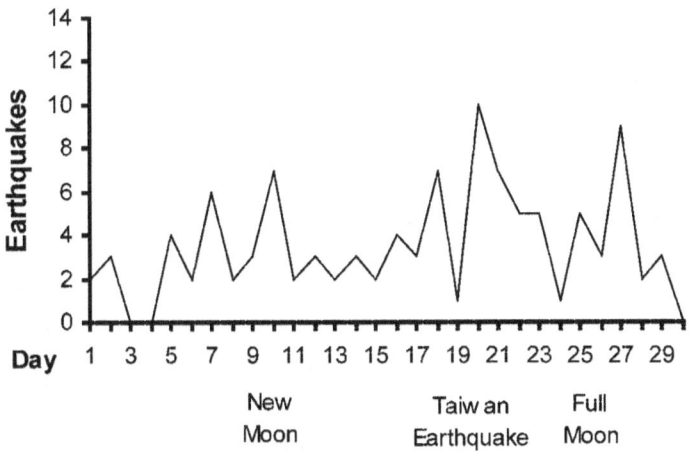

New   Taiwan   Full
Moon   Earthquake   Moon

Comparison of seismic activity

Showing numbers of earthquakes of magnitude 4.9 and above,
for the months of August and September 1999.

Figure 21

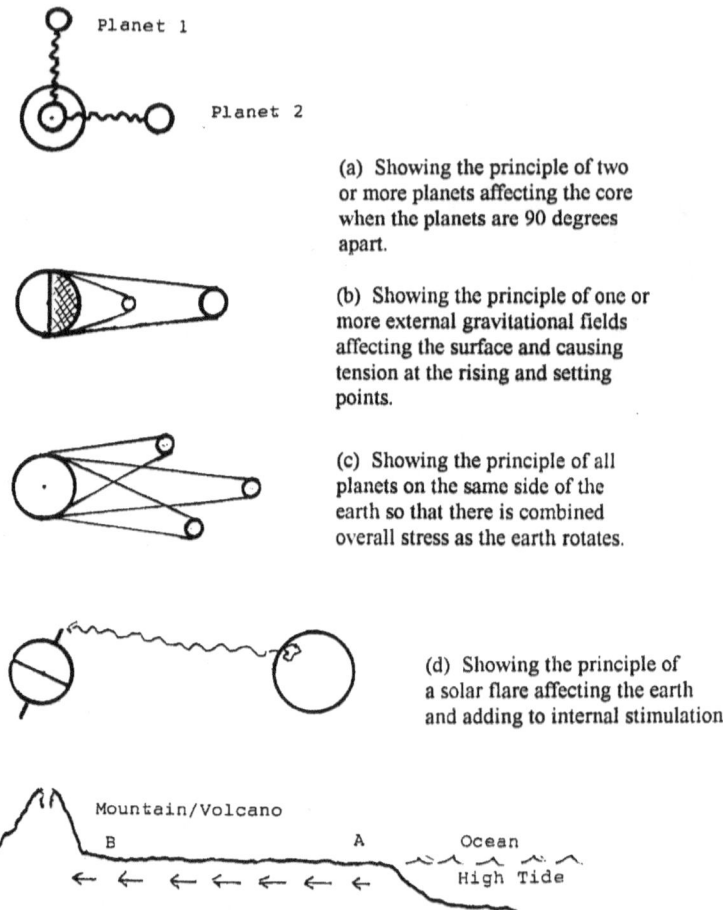

Planet 1

Planet 2

(a) Showing the principle of two or more planets affecting the core when the planets are 90 degrees apart.

(b) Showing the principle of one or more external gravitational fields affecting the surface and causing tension at the rising and setting points.

(c) Showing the principle of all planets on the same side of the earth so that there is combined overall stress as the earth rotates.

(d) Showing the principle of a solar flare affecting the earth and adding to internal stimulation.

Mountain/Volcano

B

A   Ocean

High Tide

(e) Showing the principle of a high tide helping to cause horizontal compression as the earth rotates east and the tide moves west. Conversely there is less water weight on the shelf at low tide and the shelf can spring.

Planetary and tidal effects

Showing the combination of external effects that lead to seismic and biological disturbances.

Figure 22

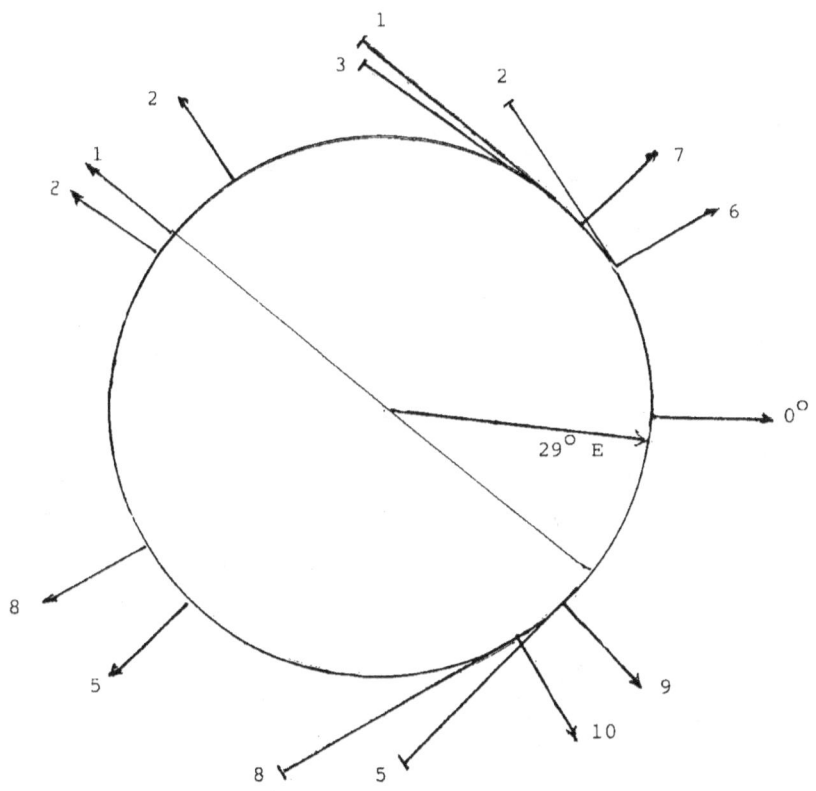

<u>lzmit, Turkey. Magnitude 7.6</u>

40 N. 29 E. 0:01 UT. August 17 1999

Showing points of Indirect Traction.
The inner arrow shows the position at 0:01 UT.

The arrow moves anti clockwise with the earth.

Number code; I Sun, 2 Mercury, 3 Venus, 5 Mars,
6 Jupiter, 7 Saturn, 8 Moon, 9 Uranus, 10 Neptune.

Figure 23

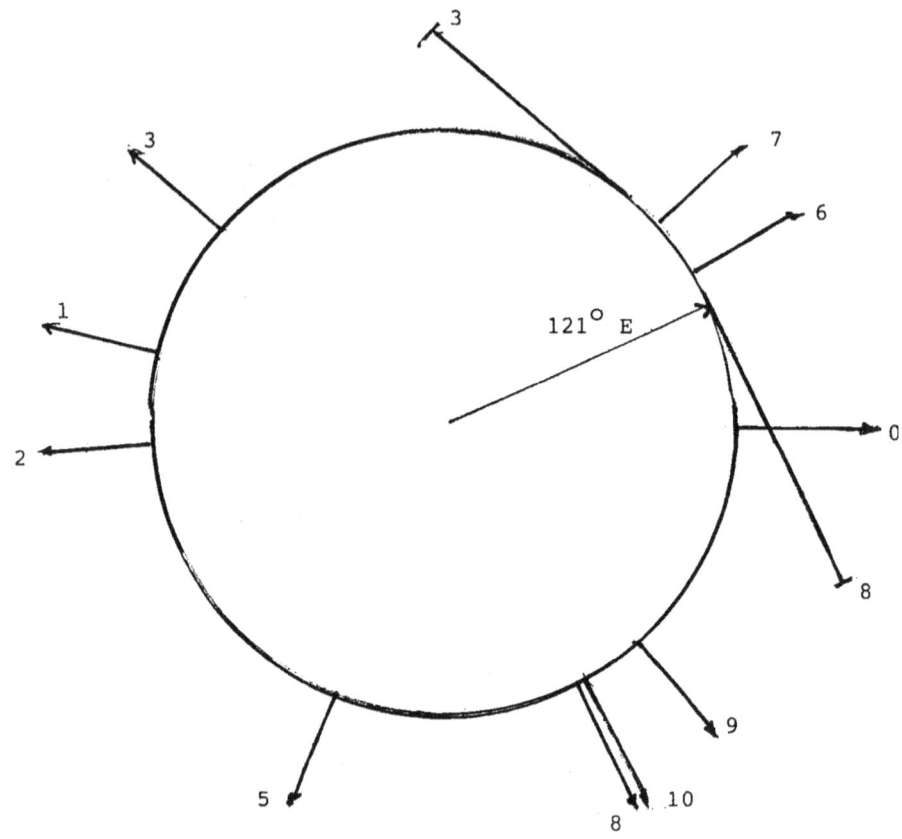

<u>Taiwan. China. Magnitude 7.7</u>

23 N, 121 E, 17:47 UT. September 20 1999
Showing points of Indirect Traction
The inner arrow shows the position at 17:47 UT.
The arrow moves anti clockwise with the earth.

Number code; 1 Sun, 2 Mercury, 3 Venus, 5 Mars,
6 Jupiter, 7 Saturn, 8 Moon, 9 Uranus, 10 Neptune.

Figure 24

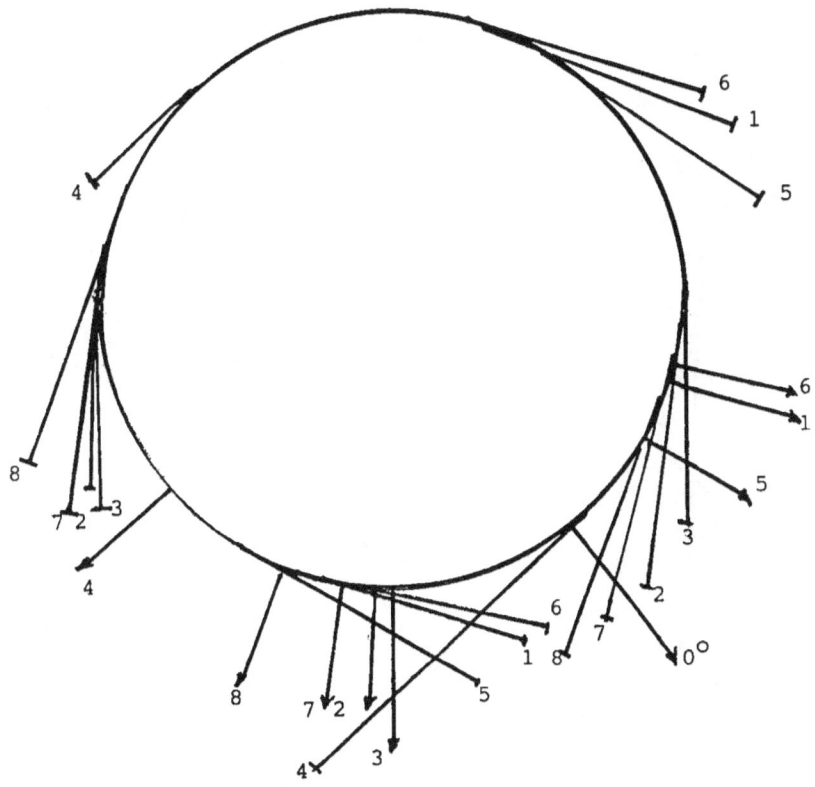

<u>Heliocentric positions of planets for August 17 1999</u>

Showing points of Indirect Traction on the sun
At the time of the Turkey Earthquake.

Code: 1 Mercury, 2 Venus, 3 Earth, 4 Mars,
5 Jupiter, 6 Saturn, 7 Uranus, 8 Neptune.

Figure 25

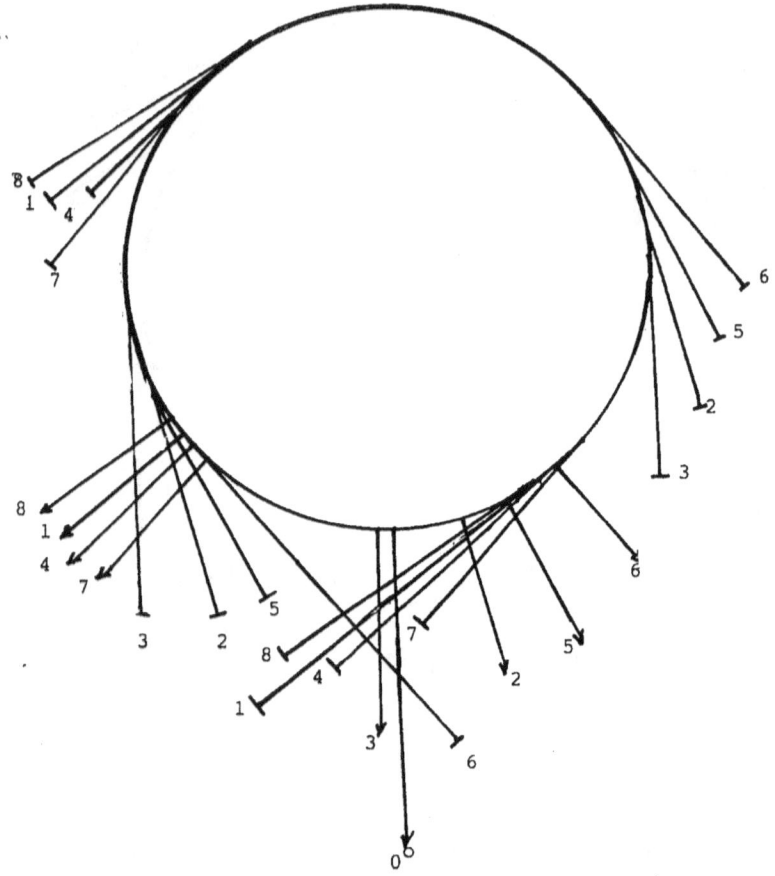

Heliocentric positions of planets for September 20 1999

Showing points of Indirect Traction on the sun
At the time of the Taiwan Earthquake

Code: 1 Mercury, 2 Venus, 3 Earth, 4 Mars,
5 Jupiter, 6 Saturn, 7 Uranus, 8 Neptune.

Figure 26

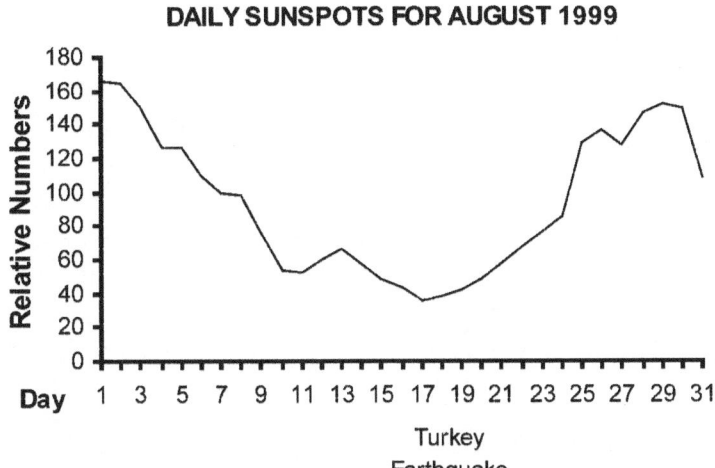

DAILY SUNSPOTS FOR AUGUST 1999

Turkey
Earthquake

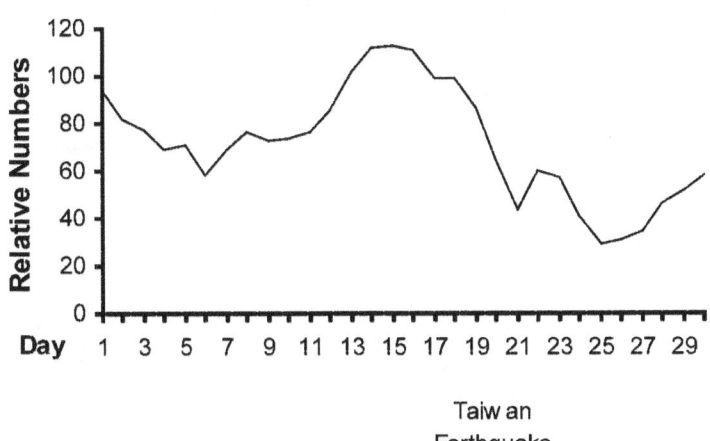

DAILY SUNSPOTS FOR SEPTEMBER 1999

Taiw an
Earthquake

Comparison of daily sunspot relative numbers

For the months of August and September 1999.

Figure 27

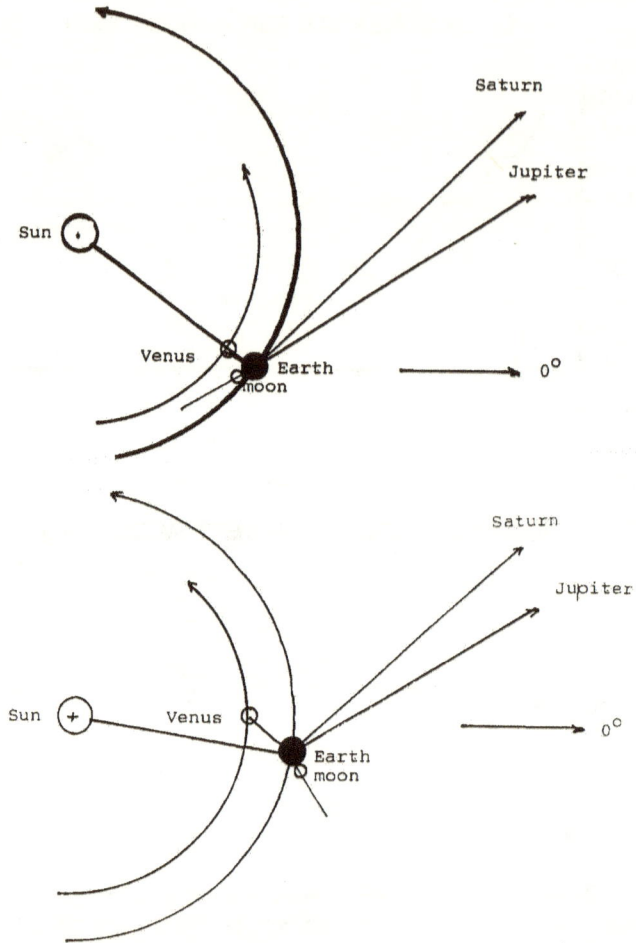

Relative positions of the earth

Top–August 17 1999, at the time of the Turkey earthquake
Bottom–September 20 1999, at the time of the Taiwan earthquake

Showing the direction and proximity of planets

Figures 28, 29

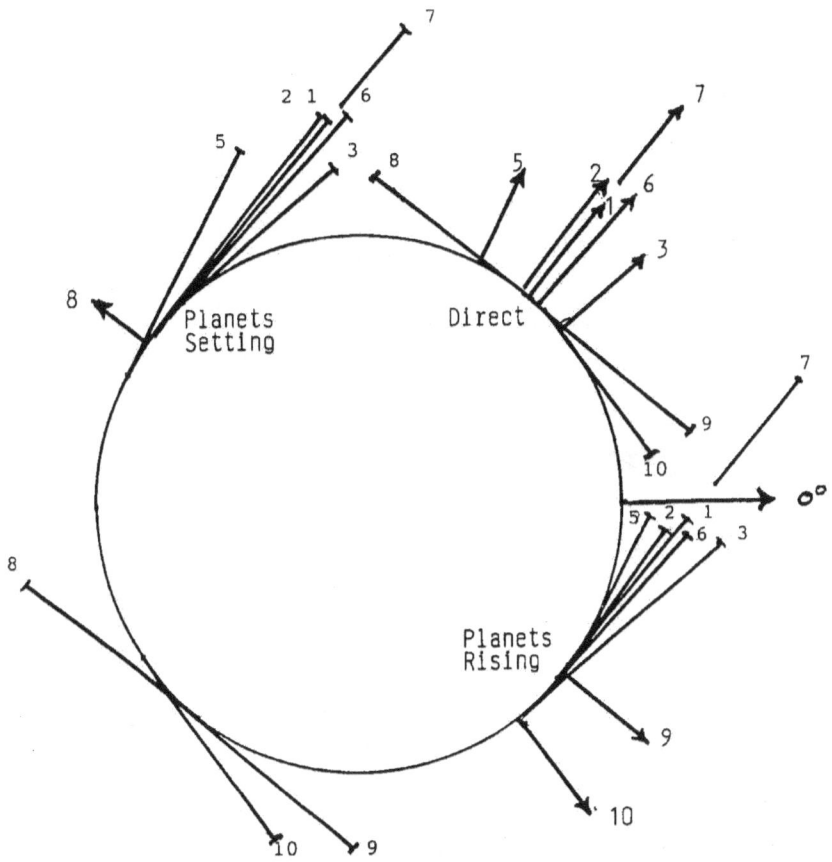

<u>Geocentric positions of planets on May 11 2000</u>

Showing grouping of planets and points of Indirect Traction

Code: 1 Sun, 2 Mercury, 3, Venus, 4 Earth, 5 Mars,
6 Jupiter, 7 Saturn, 8 Moon, 9 Uranus, 10 Neptune.

Figure 30

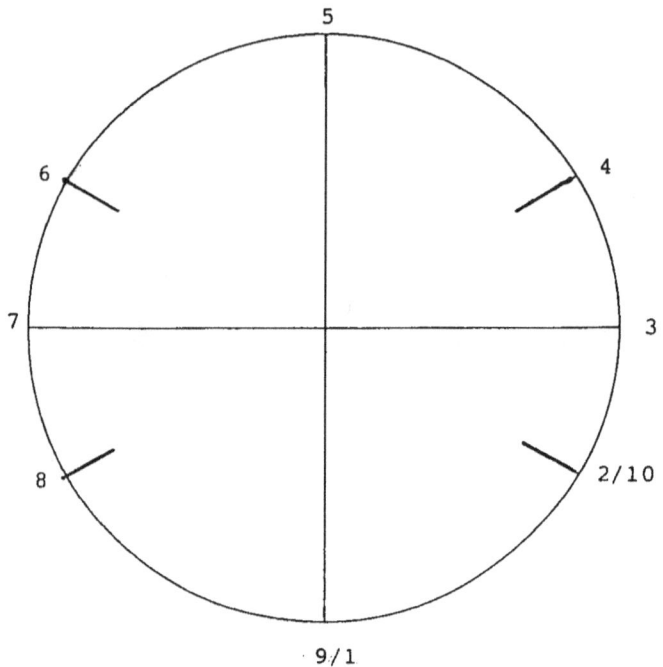

Sequence of Jupiter-Saturn alignment

1   Neutral position with no operative field between the planets.
2   Start of new cycle with a sixty-degree field between the planets.
3   Ninety-degree position with strongest planetary effect.
4   Maximum effect, with a longer field of one hundred and twenty degrees.
5   Opposition of Jupiter and Saturn. Polarity changes.
6   End of cycle causing Sunspot Minimum and start of next cycle.
7   Ninety-degree position of strongest effect.
8   An amplified sixty-degree field causing Sunspot Maximum.
9   Return to conjunction position. Polarity changes again.
10   End of cycle causing Sunspot Minimum and beginning of a new one.

Figure 31

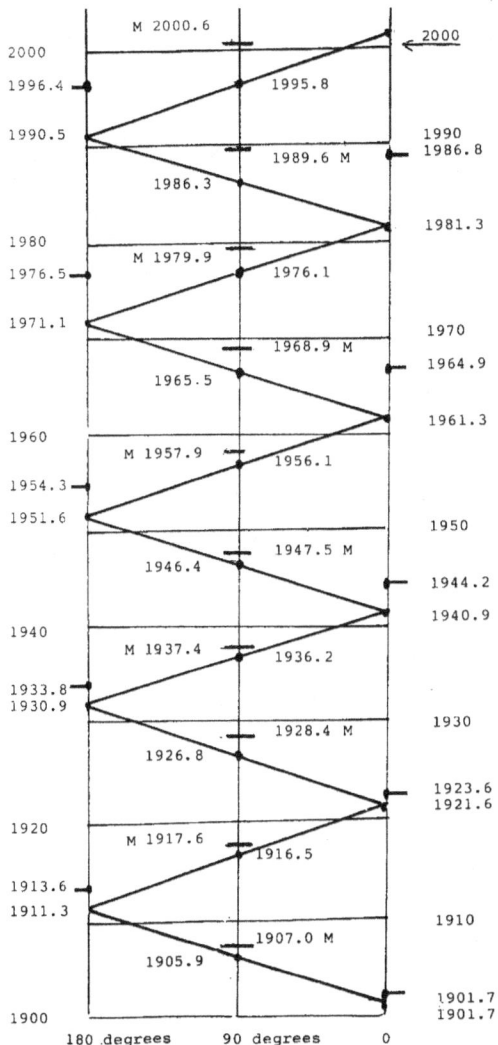

<u>Rhythms of the Jupiter-Saturn alignments and sunspot cycles 1900-2000</u>

M indicates Sunspot Maximum on the centreline,
Minimum is on the upper and lower lines

Figure 32

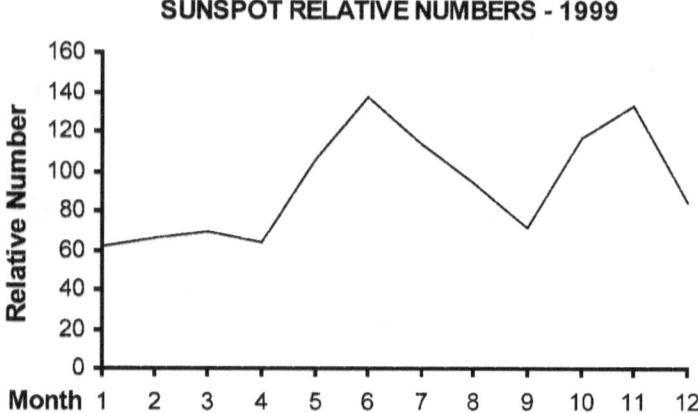

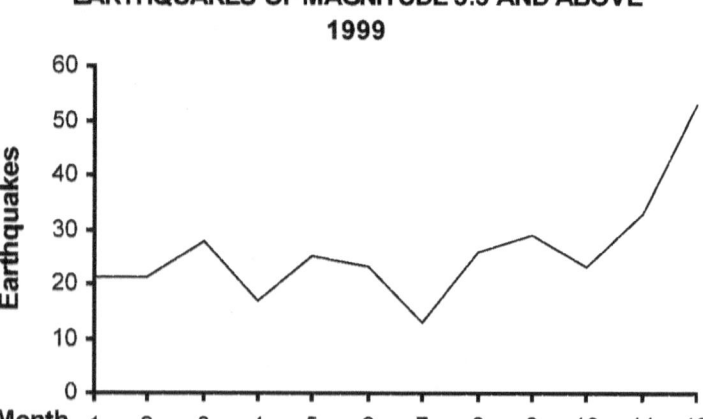

Monthly comparison of sunspots and earthquakes 5.5 and above for 1999

Figure 33

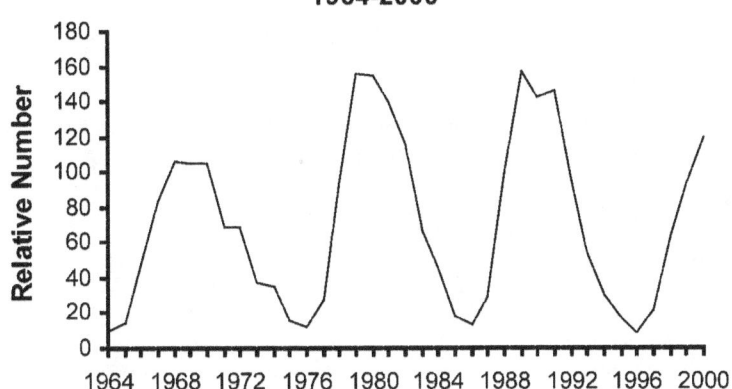

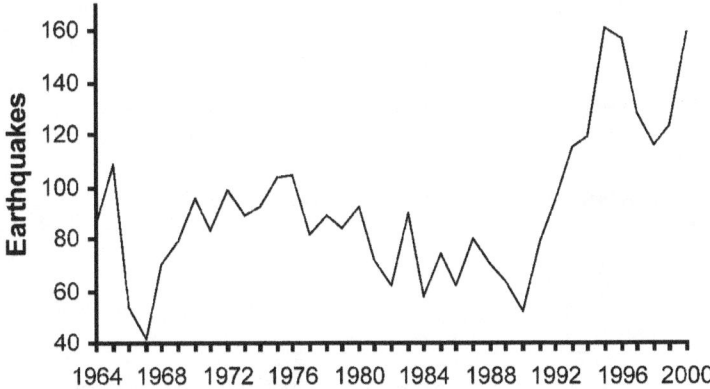

<u>Yearly comparison of sunspots and earthquakes 1964-2000</u>

Showing increases in major earthquakes after 1993.
The relevant table is in the Appendix

Figure 34

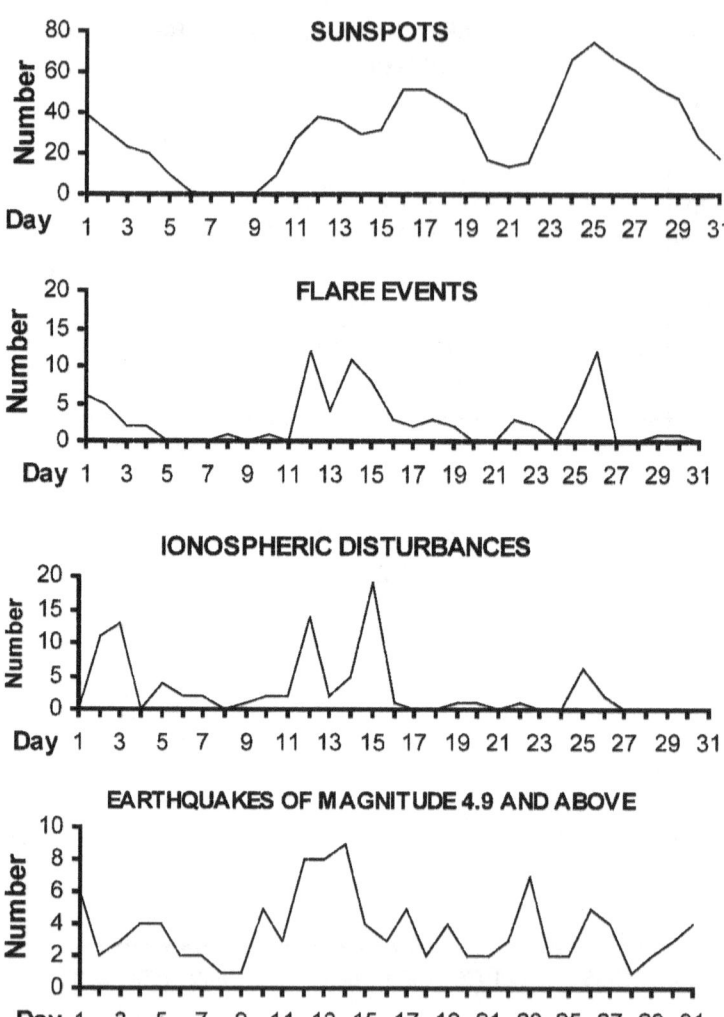

Comparison of sunspots, flare events, sudden ionospheric disturbances and seismic activity for January 1998

Figure 35

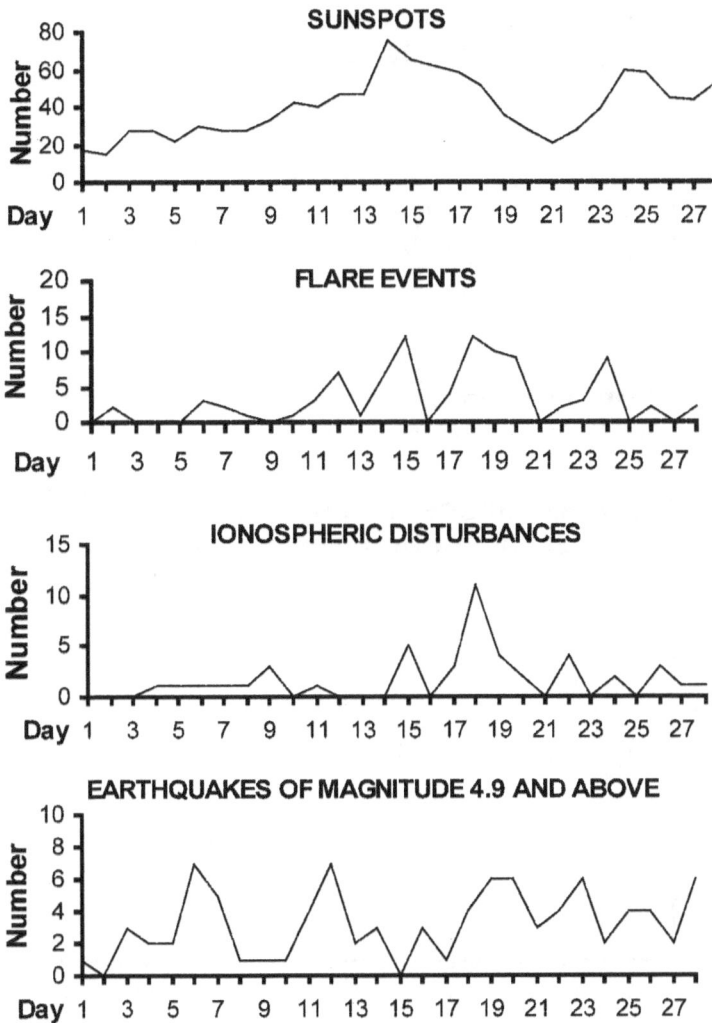

Comparison of sunspots, flare events, sudden ionospheric disturbances and seismic activity for February 1998

Figure 36

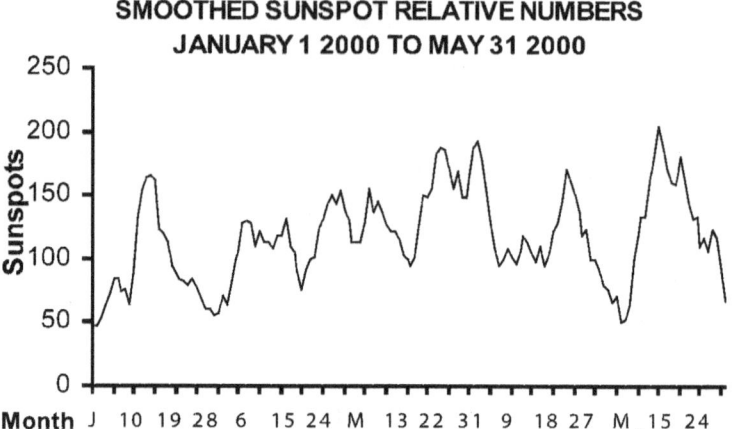

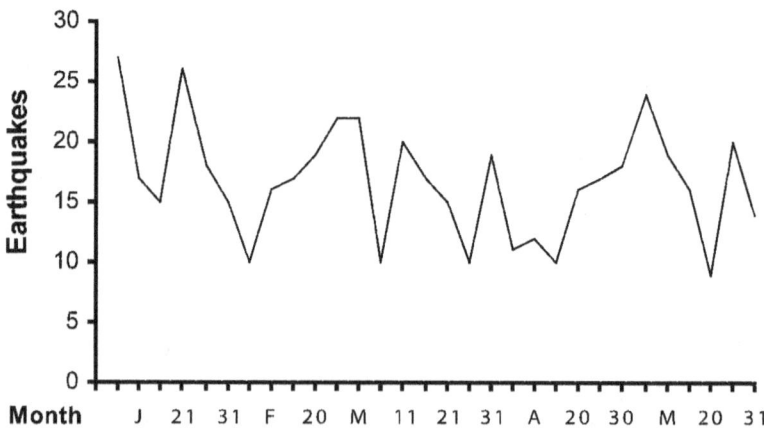

Daily comparison of sunspots and earthquakes for January–May 2000

Earthquakes are totalled for each five-day period.
The values are in the Appendix.

Figure 37

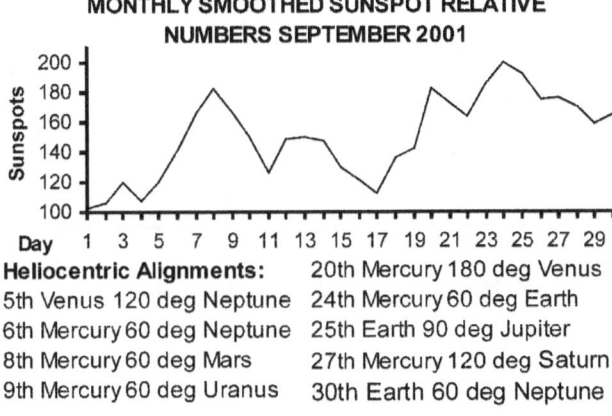

MONTHLY SMOOTHED SUNSPOT RELATIVE NUMBERS SEPTEMBER 2001

Heliocentric Alignments:
5th Venus 120 deg Neptune
6th Mercury 60 deg Neptune
8th Mercury 60 deg Mars
9th Mercury 60 deg Uranus
20th Mercury 180 deg Venus
24th Mercury 60 deg Earth
25th Earth 90 deg Jupiter
27th Mercury 120 deg Saturn
30th Earth 60 deg Neptune

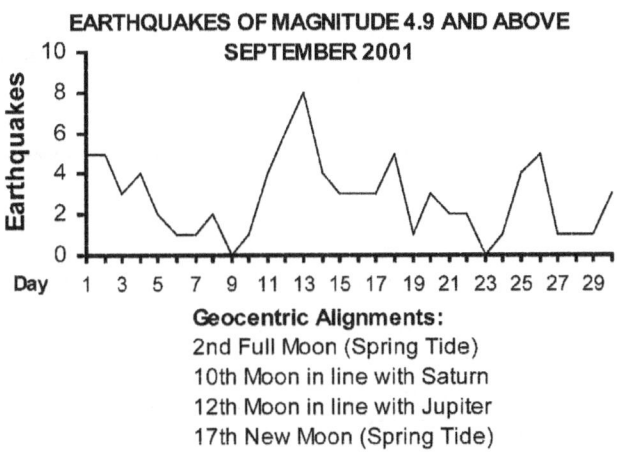

EARTHQUAKES OF MAGNITUDE 4.9 AND ABOVE SEPTEMBER 2001

Geocentric Alignments:
2nd Full Moon (Spring Tide)
10th Moon in line with Saturn
12th Moon in line with Jupiter
17th New Moon (Spring Tide)

Example of the planetary sunspot effect and the sunspot earthquake effect in
September 2001

There were no significant geocentric alignments in September: hence the lack of individual peaks in the earthquake graph, and the closer match with the sunspot graph. There is a delay of a few days from sunspot peaks to earthquake peaks.

Figure 38

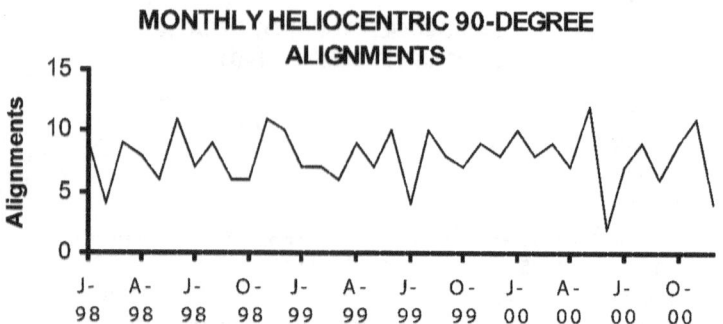

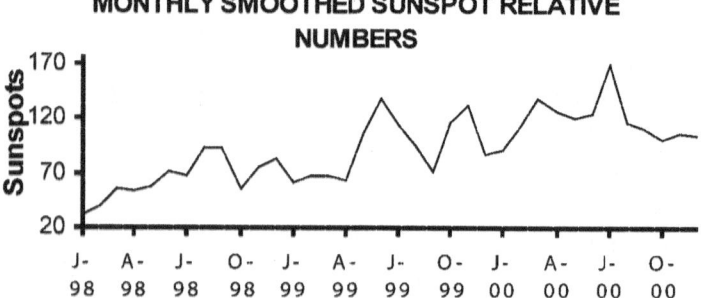

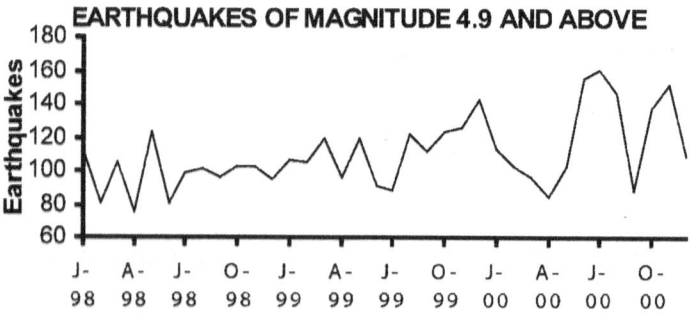

<u>Monthly heliocentric alignments with solar and seismic activity from 1998 to 2000</u>

Showing the match of heliocentric alignments with sunspot activity.

Figure 39

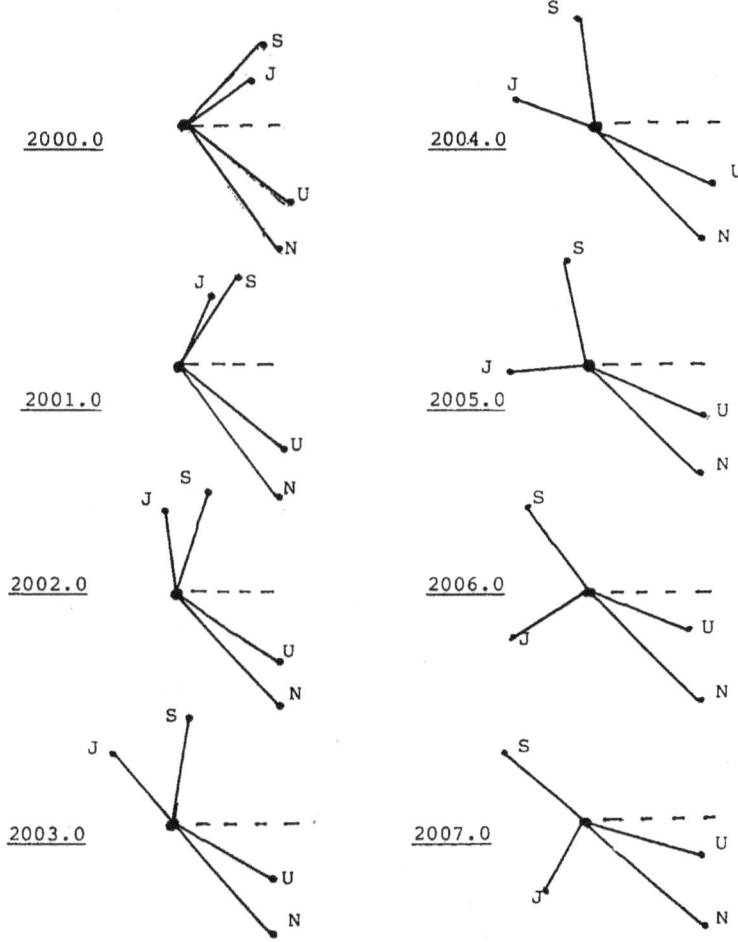

<u>Heliocentric positions of major planets</u>

First day of 2000.0 to 2007.0
Code: J-Jupiter, S-Saturn, U-Uranus, and N-Neptune.
The dotted line is zero degrees

Figure 40

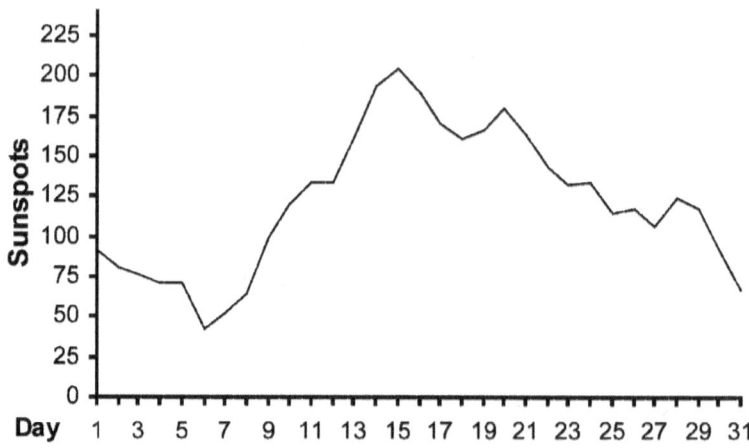

SUNSPOT RELATIVE NUMBERS AND VOLCANIC ACTIVITY FOR MAY 2000

Volcanic Activity

12th - 19th Siufrere Hills, Montserrat

13th San Cristobal, Nicaragua

14th Kavachi, Solomon Islands

15th Unimak Islands, Alaska

17th White Island, New Zealand

20th Tungurahua, Ecuador

23rd Popacatapetl, Mexico

29th Mount Cameroon, Cameroons

29th to June 6th - Krakatoa, Indonesia

A unique example of dated volcanic eruptions related to sunspot activity.

Note: May 2000 was the month of the most critical alignments. Volcanic activity and earthquakes also relate to planetary gravitational traction. Higher ocean tides also have an effect and there is not one simple cause. When extra causes coincide, or are absent, there is a clearer graph, as with this example. Source: Sunspots, STP Solar Data. Volcanoes of the World, University of North Dakota.

Figure 41

About the Author

The author has a technical background, acquired during regular service with the British Royal Air Force, and later studied philosophy. After becoming a Fellow of the Philosophical Society, he qualified as a teacher through London University. His involvement in astronomy developed from teaching in a high school where there was an active interest in the subject. He has been investigating the connection between seismic and sunspot phenomena since 1964, and since retirement in Australia has continued with his observations.

Appendix

TABLES AND LISTS OF EVENTS

1. SUNSPOT CYCLES 1886–1996
Related data for Figure 2a.

Cycle No	Year of Minimum	Year of Maximum	Sunspot Rel No's
11	1867.2	1870.6	140.5 *
12	1878.9	1883.9	74.6
13	1889.6	1894.1	87.9
14	1901.7	1907.0	64.2
15	1913.6	1917.6	105.4
16	1923.6	1928.4	78.1
17	1933.8	1937.4	119.2
18	1944.2	1947.5	151.8
19	1954.3	1957.9	201.3 *
20	1964.9	1968.9	110.6
21	1976.5	1979.9	164.5
22	1986.8	1989.6	158.5
23	1996.4	2000.5	170.1

*Uranus/Neptune, near ninety degrees.

2. JUPITER-SATURN HELIOCENTRIC ALIGNMENTS AT MIN/MAX 1900-2000
Related data for Figure 32.

Degrees apart	Year	Minimum	Maximum	Difference In years
0	1901.7	1901.7		0
90	1905.9		1907.0	1.1
180	1911.3	1913.6		2.6
90	1916.5		1917.6	1.1
0	1921.6	1923.6		2.0
90	1926.8		1928.4	1.6
180	1930.9	1933.8		2.9
90	1936.2		1937.4	1.2
0	1940.9	1944.2		3.3
90	1946.4		1947.5	1.1
180	1951.6	1954.3		2.7
90	1956.1		1957.9	1.8
0	1961.3	1964.9		3.6
90	1965.5		1968.9	3.4
180	1971.1	1976.5		5.4
90	1976.1		1979.9	3.8
0	1981.3	1986.8		5.3
90	1986.3		1989.6	3.3
180	1990.5	1996.4		5.9
90	1995.8		2000.5	4.7

Note irregularity after 1989. This may relate to the longer rhythm of Uranus and Neptune.
They were in line in 1993.3 and earlier irregularities appear to relate to the longer rhythm.

3. TIMES OF HELIOCENTRIC 90 DEGREE ALIGNMENT OF JUPITER AND SATURN AT MIN/MAX 1916 -1995

Time of Alignment	Alignment Exact Date	Sunspot Maximum	Approximate Date	Difference in years	Jupiter/Saturn angle at Max
1916.5	June 28	1917.6	August 6	1.1	68 C
1926.8	October 14	1928.4	May 25	1.1	175 O
1936.2	March 28	1937.4	May 25	1.6	69 C
1946.4	June 12	1947.5	July 1	1.1	99 O
1956.1	February 18	1957.9	November 24	1.8	61 C
1965.5	July 4	1968.9	November 24	3.4	149 O
1976.1	February 4	1978.9	November 24	3.8	21 C
1986.3	April 26	1989.6	August 6	3.3	161 O
1995.8	November 6	2000.5	July 19	4.4	5 C

Notes: "C" means the angle is closing. "O" means it is opening. There are slight differences in some dates and therefore the angles because of the interpretation of dates in decimal terms.

4. HELIOCENTRIC POSITION OF ALL PLANETS AT TIME OF JUPITER-SATURN 90 DEGREE ANGLE 1916–1995

Date	Mercury	Venus	Earth	Mars	Jupiter	Saturn	Uranus	Neptune
1916.5	351	272	275	201	19	109	317	122
1926.8	253	176	19	28	327	237	357	144
1936.2	320	308	187	39	253	343	35	225
1946.4	128	161	260	179	207	117	77	187
1956.1	215	68	148	227	146	256	120	208
1965.5	193	155	281	252	71	341	164	228
1976.1	179	231	134	107	119	29	214	251
1986.3	298	93	215	245	335	245	259	274
1995.8	185	271	43	270	263	353	299	294

Note: Various planets add to the ninety-degree effect but no regular repetitive alignments are apparent.

5. HELIOCENTRIC POSITIONS OF JUPITER AND SATURN AT MIN/MAX 1900–2000

Year	Cycle	Rel Nos	Jupiter	Saturn
1901.7	Min /14	2.6	315	296
1907.0	Max	64.2	96	345
1913.6	Min /15	1.5	285	70
1917.6	Max	105.4	56	124
1923.6	Min /16	5.6	230	200
1928.4	Max	78.1	20	255
1933.8	Min /17	3.4	185	315
1937.4	Max	119.2	287	357
1944.2	Min /18	7.7	144	86
1947.5	Max	151.8	236	131
1954.5	Min /19	3.4	98	218
1957.9	Max	201.3	195	256
1964.9	Min /20	9.6	51	330
1968.9	Max	110.6	172	23
1976.5	Min /21	12.2	43	125
1979.9	Max	164.5	148	169
1986.8	Min /22	12.3	351	250
1989.9	Max	158.5	93	285
1996.4	Min /23	8.1	279	359
2000.5	Max	170.1	42	48

Positions are to the nearest degree.

Note: The date of 2000.5 is approximate; the maximum is usually before the line up.

6. HELIOCENTRIC POSITIONS OF ALL PLANETS AT TIMES OF SUNSPOT MAXIMUM 1907–1989

Date	Mercury	Venus	Earth	Mars	Jupiter	Saturn	Uranus	Neptune
1907.0	217	118	99	187	95	345	278	101
1917.6	215	199	312	72	56	124	321	124
1928.4	165	38	243	327	20	255	4	148
1837.4	268	266	243	241	288	357	39	168
1947.5	352	55	278	33	236	131	82	190
1957.9	299	22	61	207	195	256	128	212
1968.9	217	340	61	160	172	23	180	236
1979.9	87	298	62	114	148	169	231	260
1989.6	200	213	313	162	83	281	273	281

Positions in degrees.

7. MAJOR HELIOCENTRIC ALIGNMENTS 1999-2011

Day and Year	Major Planets	Degrees apart
December 4 1999	Jupiter/Neptune	90
January 29 2000	Saturn/Uranus	90
May 7 2000	Jupiter/Uranus	90
June 22 2000	Jupiter/Saturn	0
January 17 2004	Jupiter/Saturn	60
February 28 2006	Jupiter/Saturn	90
June 27 2006	Jupiter/Neptune	90
February 2 2008	Jupiter/Saturn	120
February 23 2011	Jupiter/Saturn	180

Some of these dates were in other tables. The above table contains all the 90-degree heliocentric alignments that appear to be significant. The 60-degree alignment between Jupiter and Saturn in 2004 is hypothetically the beginning of a new cycle in the solar dynamo. The 90-degree angles in 2006 are then presumably the maximum planetary effect and the 120-degree Jupiter/Saturn alignment in 2008 is the extended effect. However, the other major alignments would modify this. In addition, when Venus, Mars or the earth, add to any 90-degree alignments the rise in sunspot activity is more noticeable, as with the high peaks at the end of 1999.

8. MAIN HELIOCENTRIC 90-DEGREE ALIGNMENTS 1998 –2000

Day and Year of Alignments	Planets making ninety-degree alignments
December 2 1998	Saturn-Neptune *
December 4 1999	Jupiter-Neptune *
January 29 2000 ..	Earth-Jupiter Saturn-Uranus *
February 1 2000 February 27 2000	Earth- Saturn Mars-Neptune
March 21st 2000	Mars-Uranus
April 24 2000	Earth-Neptune
May 7th 2000	Jupiter-Uranus *
May 8th 2000	Earth-Uranus
May 13 2000	Venus-Neptune
May 22 2000	Venus-Uranus
June 2000 **	No significant alignments and only two Mercury alignments

July 19 2000	Venus-Saturn
July 20 2000	Venus-Jupiter
August 16 2000	Earth-Saturn
September 2 2000	Venus-Neptune
September 5 2000	Venus-Mars
September 11 2000	Venus-Uranus
October 11 2000	Mars-Saturn
October 27 2000	Mars-Jupiter
October 28 2000	Earth-Neptune
October 30 2000	Earth-Venus
November 11 2000	Earth-Uranus
..	Venus-Saturn
December 25 2000	Venus-Neptune

* Indicates both are major planets ** No significant alignments
Note: All figures are to nearest date or degree. Some tables do not give
details for every day and calculations cause some slight discrepancies.
This applies to all calculated positions in this work.

9. MONTHLY PEAKS FOR SUNSPOTS AND 90-DEGREE ALIGNMENTS 1999

Year/1999 Month/No	Date of alignment	Date of Peak	Sunspot Relative no	Planets in the 90 degree Heliocentric alignment
January	9th	19th	121	Mercury – Venus
62.0	14th	..	..	Venus – Pluto
..	16th	..	..	Mercury – Mars
February	10th	12th	115	Mercury – Pluto
66.3	..	15th	144	No extra alignments
..	17th	18th	105	Venus - Neptune

March	17th	20th	78	Mercury – Venus
68.8	..	..	..	Mercury – Pluto
April	7th	9th	104	Venus – Earth
63.7	12th	..	..	Venus – Mars
May	4th	8th	152	Earth – Uranus
106.4	7th	..	..	Venus – Pluto
June	1st	6th	146	Mercury - Jupiter
137.7	2nd	..	..	Mercury - Venus
..	4th	..	..	Mercury – Saturn
..	8th	..	..	Venus – Neptune
July	22nd	30th	165	Mercury – Jupiter
113.5	28th	..	..	Mercury – Saturn
..	29th	..	..	Venus – Jupiter
August	3rd	1st	166	Earth – Saturn
93.7	28th	29th	152	Venus – Pluto
..	29th	..	..	Mercury - Jupiter
..	30th	..	152	Mercury - Jupiter
September	7th	15th	113	Mercury - Pluto
71.5	18th	..	..	Mercury – Mars
October	8th	13th	157	Venus– Uranus
116.7	25th	..	..	Mercury - Saturn
..	27th	29th	137	Earth – Neptune
November	6th	10th	206	Mercury – Venus
133.2	8th	..	..	Earth – Uranus
December	5th	18th	99	Earth – Mars
84.6	18th	..	..	Mercury – Neptune
..	..	..	..	Venus – Pluto

Note: Slower planets relate to a slower rise and fall in the relative numbers.
The table does not include all the 90-degree alignments but only for the main peaks.

10. MONTHLY HELIOCENTRIC 90-DEGREE ALIGNMENTS
1998–2000

Years	1998	1999	2000
Month	Total	Total	Total
January	9	7	10
February	4	7	8
March	9	6	9
April	8	9	7
May	6	7	12
June	11	10	2
July	7	4	7
August	9	10	9
September	6	8	6
October	6	7	9
November	11	9	11
December	10	8	4

The total numbers for each month include all exact 90-degree alignments. This includes all alignments with Mercury and Pluto. With alignments, the overall peaks do not rise during the years, as with sunspot numbers. There is therefore not an increase in alignments, which generally remain near a steady average. Slow moving planets, which help to cause the main rise in sunspots are not always exactly at 90-degrees but are still effective. Comparisons of heliocentric alignments with sunspot graphs indicate a considerable delay, which seems to vary according to the planets involved.

11. COMPARISON OF SUNSPOTS, FLARES, IONOSPHERIC DIS-TURBANCES AND EARTHQUAKES 4.9+FOR JAN/FEB 1998

1998	January				February			
Date	Sunspots	Flares	SIDS	Quakes	Sunspots	Flares	SIDS	Quakes
1	39	6	0	6	17	0	0	1
2	31	5	11	2	15	2	0	0
3	23	2	13	3	28	0	0	3
4	20	2	0	4	28	0	1	2
5	10	0	4	4	22	0	1	2
6	1	0	2	2	30	3	1	7
7	0	0	2	2	27	2	1	5
8	0	1	0	1	28	1	1	1
9	0	0	1	1	33	0	3	1
10	10	1	2	5	42	1	0	1
11	27	0	2	3	40	3	1	4
12	38	12	14	8	47	7	0	7
13	36	4	2	8	47	1	0	2
14	30	11	5	9	76	6	0	3
15	32	8	19	4	65	12	5	0
16	52	3	1	3	62	0	0	3
17	52	2	0	5	58	4	3	1
18	46	3	0	2	52	12	11	4
19	39	2	1	4	35	10	4	6
20	17	0	1	2	28	9	2	6
21	14	0	0	2	21	0	0	3
22	16	3	1	3	28	2	4	4
23	40	2	0	7	39	3	0	6
24	66	0	0	2	59	9	2	2
25	75	5	6	2	58	0	0	4
26	67	12	2	5	45	2	3	4
27	61	0	0	4	43	0	1	2
28	53	0	0	1	54	2	1	6
29	47	1	0	2	-	-	-	-
30	28	1	0	3	-	-	-	-
31	18	0	0	4	-	-	-	-

Sunspot figures are daily Sunspot Relative Numbers
Flares are all daily flare events
SIDS are daily Sudden Ionospheric Disturbances
Earthquakes are all earthquakes of magnitude 4.9 and above.

12. DAILY SUNSPOT RELATIVE NUMBERS 1999
General data relating to graphs.

Day	Jan	Feb	Mar	Apr	May	June	July	Aug	Sept	Oct	Nov	Dec
1	64	29	77	44	76	137	141	168	94	50	115	103
2	68	25	88	39	92	135	146	165	82	64	90	99
3	58	19	102	48	80	131	142	151	77	68	81	70
4	65	15	96	71	84	140	134	127	69	77	86	57
5	64	12	99	81	89	144	130	127	71	124	102	51
6	48	19	79	92	105	146	122	110	58	136	103	63
7	64	28	43	82	141	123	117	100	69	138	123	59
8	51	36	41	89	151	131	113	98	76	113	146	78
9	47	41	63	104	149	160	115	76	73	131	169	87
10	46	60	61	90	136	146	112	54	74	130	206	110
11	32	78	76	75	134	159	115	52	76	126	205	101
12	35	115	87	76	122	153	130	60	85	145	188	89
13	41	134	76	81	101	152	103	67	102	157	164	102
14	65	138	94	74	93	147	84	57	112	151	146	104
15	83	144	91	63	105	139	80	49	113	115	153	92
16	83	133	97	67	105	120	77	44	111	122	171	94
17	93	122	99	75	99	97	79	36	99	125	166	102
18	102	105	96	55	93	80	90	38	99	114	159	99
19	121	85	101	50	94	79	79	42	86	109	164	100
20	120	86	78	50	96	62	77	48	65	113	152	102
21	114	74	71	42	109	79	79	58	44	97	142	93
22	108	47	61	40	97	106	94	68	60	86	137	94
23	87	42	30	42	86	144	97	76	57	79	110	89
24	68	38	41	51	85	195	113	86	41	90	103	97
25	31	44	37	45	92	194	119	129	31	120	102	93
26	30	51	31	47	114	182	100	136	29	143	95	84
27	28	59	24	55	119	172	115	128	35	140	124	69
28	28	77	37	55	119	169	144	147	46	135	105	62
29	24	-	51	61	115	160	161	152	52	137	78	75
30	24	-	55	66	109	148	165	150	58	155	93	48
31	30	-	51	-	109	-	146	109	-	129	-	57
Mean	62.0	66.3	68.8	63.7	106.4	137.7	113.5	93.7	71.5	116.7	133.2	84.6

Solar data IPS Space Services

13. DAILY EARTHQUAKES MAGNITUDE 4.9+1999
General data relating to graphs.

Day	Jan	Feb	Mar	Apr	May	Jun	Jul	Aug	Sept	Oct	Nov	Dec
1	2 F	6	5	4	0	7	4	8	2	3	4	6
2	2	2	6 F	4	4	3	2	3	3	5	3	6
3	1	3	1	4	3	3	3	6	0	3	5	2
4	3	7	12	3	0	2	2	2	0	6	0	7
5	5	4	3	2	5	1	2	6	4	6	3	5
6	1	3	3	9	8	3	0	1	2	1	2	3
7	5	1	5	3	5	3	7	3	6	5	1	7 N
8	1	1	4	2	6	1	3	2	2	5	7 N	5
9	3	3	5	2	4	8	5	2	3 N	1 N	6	4
10	1	2	3	0	4	6	0	6 N	7	6	1	5
11	1	4	4	2	4	1	4	3	5	5	3	6
12	7	3	5	2	3	3	3	5	3	5	8	6
13	1	2	1	4	1	2 N	6 N	9	2	3	5	2
14	3	7	3	2	4	0	1	2	3	2	0	1
15	2	4	1	2	2 N	3	6	4	2	1	6	4
16	6	3 N	4	3 N	3	3	2	3	4	9	4	5
17	4 N	4	0 N	4	5	3	0	12	3	4	5	5
18	5	4	7	2	10	5	3	4	7	6	5	2
19	4	3	1	2	2	1	4	2	1	2	8	5
20	3	6	3	4	5	1	4	4	10	6	2	4
21	0	1	6	1	5	5	3	3	7	5	6	5
22	2	4	2	5	5	5	4	6	5	5	2	6 F
23	1	8	5	4	1	2	4	1	5	3	3 F	5
24	8	6	1	3	5	2	2	3	1	5 F	2	3
25	8	8	4	3	6	2	1	2	5 F	5	4	6
26	7	3	4	3	4	3	5	10 F	3	2	14	3
27	5	3	4	1	4	3	3	0	9	4	6	1
28	5	0	7	7	1	1 F	3 F	4	2	1	4	4
29	3	0	4	4	5	6	1	2	3	1	6	9
30	4	0	3	5 F	4 F	3	3	0	0	5	0	9
31	4 F	--	3 F	--	1	--	3	3	--	3	--	2
Total	107	105	119	96	119	91	89	122	111	123	126	143

NCEDC Geo Survey Berkeley

N = New Moon. F = Full Moon

14. DAILY SUNSPOT RELATIVE NUMBERS 2000

Day	Jan	Feb	Mar	Apr	May	June	July	Aug	Sep	Oct	Nov	Dec
1	48	71	138	187	91	85	145	106	142	115	140	116
2	51	64	130	193	80	79	141	110	118	164	147	109
3	54	81	114	177	76	75	124	107	128	153	141	118
4	64	99	113	164	71	101	114	110	134	150	130	72
5	73	104	113	129	71	95	127	144	114	128	133	65
6	85	136	136	108	42	99	154	143	114	97	108	57
7	85	130	155	94	52	105	177	164	110	66	122	68
8	75	128	145	100	64	120	177	140	85	72	127	57
9	76	109	146	108	99	122	179	128	63	71	95	58
10	65	122	137	102	120	119	215	154	42	57	101	62
11	90	114	127	96	133	151	202	165	26	82	90	72
12	134	113	122	113	133	147	186	170	35	122	72	89
13	153	108	121	118	161	156	194	176	55	121	70	114
14	164	119	115	114	193	171	164	204	60	104	84	135
15	157	118	103	105	205	158	148	183	77	83	98	153
16	163	131	100	98	189	142	197	178	85	92	95	145
17	131	109	95	110	170	139	224	152	108	97	94	151
18	120	104	101	94	161	147	228	140	112	95	116	138
19	114	89	126	103	167	145	246	133	121	90	125	118
20	95	76	150	121	180	159	241	106	124	94	110	127
21	88	92	148	128	163	147	231	77	137	97	120	116
22	84	100	156	145	143	127	216	67	142	89	113	107
23	82	95	182	170	132	124	199	67	160	85	91	102
24	80	123	188	160	134	119	171	77	163	82	98	115
25	85	131	185	151	115	111	177	81	153	88	74	108
26	77	144	170	136	117	129	133	79	161	73	59	121
27	70	150	155	118	106	138	126	113	162	80	84	118
28	60	151	169	124	124	115	120	132	142	106	106	118
29	61	162	148	100	117	109	113	138	119	113	123	100
30	51	-	148	100	93	114	112	144	100	108	138	111
31	58	-	164	-	67	-	93	157	-	111	-	87
Mean	90.1	112.9	138.5	125.5	121.6	124.9	170.1	130.5	109.7	99.4	106.8	104.4

Solar Data IPS Space Services

There were two days in November 1999 with sunspot numbers above 200 and one day in May 2000. The highest monthly number before June was 138.2 in March 2000. The highest above is 246 on July 19.

15. DAILY EARTHQUAKES 4.9+2000

Day	Jan	Feb	Mar	Apr	May	June	July	Aug	Sep	Oct	Nov	Dec
1	4	1	5	1	4	1	3	10	2	3	5	8
2	5	2	1	2	3	2	4	5	4	5	1	2
3	1	4	3	5	2	4	2	11	4	5	3	2
4	1	1	2	1	10	11	6	9	2	7	3	8
5	11	2	3	2	5	9	1	3	4	4	4	4
6	5	2	1	3	2	10	9	3	1	5	3	5
7	1	2	3	5	2	4	4	5	0	4	7	6
8	8	1	7	3	8	9	3	2	4	6	3	5
9	4	4	3	0	2	11	1	3	3	2	5	5
10	1	7	5	1	5	10	5	1	11	2	3	5
11	3	1	2	4	1	4	9	2	2	4	0	6
12	1	6	3	0	5	2	8	3	4	3	1	4
13	2	2	0	3	3	1	6	3	2	1	5	12
14	5	6	4	2	4	10	7	4	3	7	2	4
15	3	2	3	1	3	6	7	12	1	1	3	7
16	4	4	7	1	0	7	8	1	3	6	30	3
17	5	3	3	5	4	5	5	6	3	3	9	5
18	5	3	1	4	1	3	2	5	2	8	12	1
19	3	7	3	3	2	6	8	2	3	1	6	4
20	7	2	4	3	2	3	2	3	3	1	1	10
21	6	9	4	3	3	6	9	1	2	11	9	8
22	5	6	4	2	4	5	2	3	4	5	6	7
23	7	2	2	3	5	3	5	6	4	3	10	5
24	0	1	1	3	5	1	1	2	0	4	5	5
25	1	4	2	6	3	2	7	0	2	6	3	3
26	5	5	1	7	5	2	7	3	3	4	5	2
27	2	7	3	0	3	2	6	3	3	6	1	2
28	7	2	8	1	2	5	6	8	7	5	0	6
29	2	3	4	4	0	10	5	3	1	7	4	3
30	2	-	1	6	2	1	6	5	1	11	2	4
31	2	-	3	-	2	-	8	9	-	4	-	5
Total	118	102	96	84	102	155	161	146	88	138	151	154

Earthquakes CNSS catalog Berkeley

16. SMOOTHED YEARLY SUNSPOT RELATIVE NUMBERS
1960–2000

Year	Rel Nos	Year	Rel Nos	Year	Rel Nos	Year	Rel Nos
1960	112.3	1970	104.5	1980	154.6	1990	142.6
61	53.9	71	68.6	81	140.4	91	145.7
62	37.6	72	68.9	82	115.9	92	94.3
63	27.9	73	38.0	83	66.6	93	54.6
64	10.2	74	34.5	84	45.9	94	29.9
65	15.1	75	15.5	85	17.9	95	17.5
66	47.0	76	12.6	86	13.4	96	8.6
67	83.8	77	27.5	87	29.4	97	21.5
68	105.9	78	92.5	88	100.2	98	64.3
69	104.5	79	155.4	89	157.6	99	93.3
						2000	119.6

17. TOTAL EARTHQUAKES 5.5+FROM SUNSPOT MINIMUM
1964–2000

Yr	Magnitude 5.5-5.9	6+	Total EQ	Yr	Magnitude 5.5-5.9	6+	Total EQ	Yr	Magnitude 5.5-5.9	6+	Total EQ	Yr	Magnitude 5.5-5.9	6+	Total EQ
60				70	253	96	349	80	234	92	326	90	289	52	341
61				71	276	83	359	81	204	72	276	91	231	79	310
62				72	293	99	392	82	239	62	301	92	303	95	398
63				73	289	89	378	83	305	90	395	93	247	115	362
64	273	86	359	74	295	92	387	84	263	58	321	94	287	119	406
65	410	109	519	75	274	104	378	85	247	74	321	95	331	161	492
66	223	54	277	76	329	105	434	86	263	62	325	96	321	157	478
67	200	42	242	77	319	82	401	87	268	80	348	97	287	128	415
68	261	70	331	78	292	89	381	88	254	70	324	98	236	116	352
69	251	79	330	79	244	84	328	89	241	64	305	99	235	123	358
												2000	218	159	377

Sunspot Relative Numbers–IPS Space Services, Earthquake data
NCEDC Berkeley, California

Note: In 1993 the percentage of disturbances over magnitude 6.0 rose from an average of 25% to over 305. (1999 had 34%).

18. MONTHLY SUNSPOTS AND EARTHQUAKES 4.9+1996 -2000

	S'spot 1996	EQ 1996	S'spot 1997	EQ 1997	S'spot 1998	EQ 1998	S'spot 1999	EQ 1999	S'spot 2000	EQ 2000
Jan	11.5	109	5.7	112	31.9	111	62.0	107	90.1	113
Feb	4.4	149	7.6	80	40.3	81	66.3	105	112.9	103
Mar	9.2	116	8.7	84	54.8	105	68.8	119	138.5	96
Apr	4.8	85	15.5	140	53.4	75	63.7	96	125.5	84
May	5.5	59	18.5	126	56.3	123	106.4	119	121.6	102
June	11.8	147	12.7	97	70.7	81	137.7	91	124.9	155
July	8.8	95	10.4	111	66.6	99	113.5	89	170.1	161
Aug	14.4	133	24.4	88	92.2	101	93.7	122	130.5	146
Sept	1.6	212	51.3	97	92.9	96	71.5	111	107.7	88
Oct	0.9	150	22.8	132	55.5	103	116.7	123	99.4	138
Nov	17.9	193	39.0	127	74.0	102	133.2	126	106.8	151
Dec	13.3	104	41.2	192	81.9	95	84.6	143	104.5	109

Earthquakes CNSS Berkeley/Sunspots STP Solar Data Internet

19. ANNUAL HELIOCENTRIC POSITIONS OF PLANETS 2000-2007
(For Figure 40 as at January 1st of each year)

Planets	*2000.0*	*2001.0*	*2002.0*	*2003.0*	*2004.0*	*2005.0*	*2006.0*	*2007.0*
Mercury	252	292	338	43	124	190	232	268
Venus	181	46	272	137	1	229	92	317
Earth	99	100	100	100	99	100	100	100
Mars	359	182	26	201	51	222	74	244
Jupiter	36	69	100	130	159	187	214	242
Saturn	45	59	72	86	99	113	127	140
Uranus	316	320	324	328	332	336	340	344
Neptune	303	306	308	310	312	314	317	319
Pluto	250	252	355	257	259	262	264	266

The positions are in degrees from zero degrees Aries, counting anti-clockwise.

20. MAJOR EARTHQUAKES MAGNITUDE 6+1999
(Universal Time. Latitude minus is South, Longitude minus is West)

Date	*Time*	*Latitude*	*Longitude*	*Magnitude*	*Area*
January					
12	2:32	26. 7	140.1	6.1	
19	3:35	-4. 5	153.2	7.0	Solomon Islands
24	0:37	30. 6	131.0	6.4	
24	7:01	-21. 1	-174.6	6.4	
25	18:19	4. 4	-75.7	6.2	
28	8:10	52. 8	-169.1	6.6	
28	18:24	-4. 5	153.6	6.4	New Guinea/Solomon
February					
2	1:13	-20. 3	-174.3	6.3	
6	21:47	-12. 8	166.6	7.4	Samoa Islands
13	14:45	-3. 5	144.8	6.3	New Guinea
22	1:00	-21. 4	169.6	6.5	
22	13:49	24. 1	122.6	6.1	
25	18:58	51. 6	104.8	6.1	New Guinea

March					
2	7:'2	35. 5	141.7	6.1	
3	17:45	-22. 7	-68.5	6.1	
4	5:38	31. 3	57.1	6.6	
8	12:25	52. 0	159.5	6.9	
18	17:55	41. 0	142.9	6.1	
20	10:47	51. 58	-177.6	6.9	
28	19:05	30. 5	79.4	6.6	
31	5:54	5. 8	-82.6	6.8	
April					
1	21:36	-4. 3	152.7	6.3	Solomon Islands
2	17:05	-19. 8	168.1	6.2	
3	6:17	-16. 6	-72.6	6.8	
6	8:22	-6. 5	147.0	6.3	New Guinea
9	12:16	-26. 3	178.2	6.2	
11	16:50	-6. 0	148.4	6.1	E. Solomon Islands
13	10:38	-21. 4	-176.4	6.8	
20	19:04	-13. 8	-179.0	6.5	
May					
5	22:41	14. 3	-94. 6	6.3	
6	23:00	29. 5	51. 8	6.2	
8	22:12	14. 2	97. 9	6.3	
10	20:23	-5. 1	150. 8	7.1	E. Solomon Islands
12	17:59	43. 0	143. 8	6.2	Solomon Islands
16	0.51	-4. 7	152. 4	7.1	E. Solomon Islands
16	15:25	-2. 6	138. 2	6.4	W. New Guinea
17	10.07	-5. 1	152. 8	6.7	E. Solomon Islands
22	10:08	-20. 7	169. 8	6.1	
June					
6	7:08	13.8	-90.7	6.3	
18	10:55	5.5	126.6	6.4	S. Philippines
21	17:43	18.3	-101.5	6.3	
July					
3	5:30	26.3	140.8	6.3	
9	5:04	-6.5	154.9	6.3	Solomon Islands
11	11.51	15.7	-88.3	6.7	

19	2:17	-28.6	-177.6	6.4	
26	1;33	-5.15	151.9	6.2	New Guinea
28	0:16	-28.6	-177.5	6.1	
28	10:08	-30.2	-178.0	6.3	
August					
1	8:39	-30.3	-177.8	6.6	
12	5:45	-1.7	123.4	6.1	
17	0:01	40.7	29.9	7.6	Turkey
21	21:51	-58.2	-13.0	6.3	
22	9:35	-40.4	-74.8	6.4	
22	12:40	-16.1	168.0	6.5	
26	1:24	10.4	126.1	6.1	
26	7:39	-3.5	145.8	6.3	New Guinea
28	12:40	-1.3	-77.5	6.0	
September					
15	3:00	-20.9	-67.2	6.4	
17	14:54	-13.7	167.1	6.3	
18	21:28	51.1	157.5	6.2	
20	17:47	23.7	121.0	7.7	Taiwan
20	17:57	23.5	121.2	6.0	..
20	18:03	23.7	121.3	6.3	..
20	18:11	23.3	121.2	6.1	..
22	0:14	23.6	121.0	6.5	..
22	0:49	23.6	121.1	6.4	..
25	23:52	23.7	121. 1	6.5	..
28	5:00	54.8	167.9	6.1	
October					
10	7:03	-1.9	134.2	6.1	Ambon, Indonesia
13	1:33	54.6	-161.1	6.5	
18	2:43	56.1	-26.5	6.2	
23	2:12	-4.7	153.3	6.2	New Britain
24	4:21	44.5	149.5	6.0	
November					
1	17:53	-1.9	134.2	6.3	
8	11:45	36.5	71.2	6.5	
11	2:41	49.3	155.6	6.1	
15	5:42	-1.3	88.9	7.0	

17	3:27	-5.9	148.8	7.0	New Guinea
17	11:36	-6.0	148.6	6.3	..
19	13:56	-6.3	148.8	7.0	..
21	3:51	-21.8	-68.8	6.0	
21	6:46	18.3	-107.2	6.2	
26	13:21	-16.4	168.2	7.5	
29	3:46	-1.4	88.9	6.5	
December					
1	19:23	17.8	-82.4	6.3	
7	0:19	57.41	-154.4	6.4	
7	21;29	-15.8	-174.0	6.4	
8	13:34	-9.8	159.9	6.2	
9	10:18	-6.0	148.1	6.4	New Guinea
10	18:38	-36.2	-97.1	6.5	
15	4:41	-5.7	150.8	6.2	New Guinea
16	14:17	-50.2	139.8	6.0	
17	0:27	-50.2	139.5	6.1	
17	4:03	7.9	-38.0	6.0	
18	17:44	-2.4	139.6	6.4	New Guinea
19	0:48	13.0	144.5	6.0	
21	14:14	-6.8	105.6	6.6	Malaysia
24	19:26	-56.1	146.9	6.1	
29	5:19	18.1	-101.3	6.0	
29	13:29	-10.9	165.3	6.8	
29	19:15	-10.8	105.2	6.3	Malaysia
29	22:53	-11.0	165.2	6.2	W of New Guinea

NCEDC Geo Survey Berkeley

Note: The areas named are all near the edge of the Pacific Plate. Corresponding human disturbance occurred in Indonesia before and after the earthquakes.

21. CONVERSION TABLE–RIGHT ASCENSION TO DEGREES

R.A. Hours	Degrees	Area	Decimals - Days		
00:00	0	0 Aries	0.1	=	36.5
01:00	15		0.2	=	73.0
02:00	30	0 Taurus	0.3	=	109.5
03:00	45		0.4	=	146.0
04:00	60	0 Gemini	0.5	=	182.5
05:00	75		0.6	=	219.0
06:00	90	0 Cancer	0.7	=	255.5
07:00	105		0.8	=	292.0
08:00	120	0 Leo	0.9	=	328.5
09:00	135		1.0	=	365.0
10:00	150	0 Virgo	*************		
11:00	165		*Decimals - Date*		
	180	0 Libra	0.1	=	Feb 5
	195		0.2	=	Mar 14
14:00	210	0 Scorpio	0.3	=	Apr 19
15:00	225		0.4	=	May 21
16:00	240	0 Sagittarius	0.5	=	July 1
17:00	255		0.6	=	Aug 6
18:00	270	0 Capricorn	0.7	=	Sep 12
19:00	285		0.8	=	Oct 19
20:00	300	0 Aquarius	0.9	=	Nov 24
21:00	315		1.0	=	Dec 31
22:00	330	0 Pisces	*************		
23:00	345		1 degree = 4 minutes		
24:00	360	0 Aries			

GLOSSARY

Alignments
The positions and relationships of planets to each other.

Aphelion
A point when a planet is at its most remote distance from the sun.

Apogee
A point when the moon is at its greatest distance from the earth.

Aries
A traditional point of celestial coordinates from which other measurements can be calculated, such as Right Ascension in hours and minutes, or in degrees for navigation.

Armature
A rotating part of an electric generator that induces a voltage as it sweeps through a magnetic field.

Aseismic Plates
Areas of the earth's crust above the asthenosphere in which there are fewer earthquakes.

Asteroid Belt
An area between the orbit of Mars and Jupiter where there are various masses of rock on different orbits.

Asthenosphere
An area in the earth's upper mantle, situated beneath the crust. It ranges from 50 to 250 kilometres in depth.

Astronomical Unit.	A measurement of distance based on the mean distance of the earth from the sun. (Approximately 93 million miles or 150 million kilometres.)
Aurora Borealis	The light phenomenon created by electrically charged particles at the north pole, called the northern lights. The southern lights are Borealis Australis.
Celestial Latitude	The angular position of a celestial body above or below the plane of the ecliptic.
Celestial Longitude	The longitudinal definition of the position of a celestial body.
Conjunction	An alignment when two celestial bodies are in the same longitude.
Cosmic Rays	High energy charged particles that travel at near the speed of light.
Crust	The outer area of the earth, varying from 5 to 60 kilometres in thickness.
Declination	The angular measurement of a celestial body north or south of the celestial equator, or ecliptic.
Direct Traction	Gravitational force acting in a direct line of force from the celestial body exerting the force.
Earth Tide	A movement of the crust of the earth, caused by external gravitational force, as from the sun and the moon.

Ecliptic	The plane of the orbit of the earth round the sun from which celestial latitude is determined.
Electromagnetic Energy	Energy waves caused by the acceleration of charged particles.
Equinox	Times when the sun crosses the equator, caused by the inclination of the earth on its axis in relation to its position on its orbit.
Geocentric Coordinates	The alignments of celestial bodies with the earth as the centre of focus, usually measured in hours and minutes but often in degrees for navigation.
Gravitational field	The traction zone of an external body applying to part of the earth.
Greenwich Mean Time	A universal time system based on the position of zero degrees in longitude in relation to zero degrees of celestial longitude.
Heliocentric Coordinates	The alignments of planets with the sun as the centre or focus, measured anti-clockwise, in degrees.
Indirect Traction	This is the gravitational force, which is at a tangent to the direct line of force. It is to a point on the earth where the external body is either rising or just setting.
Ion	An electrically charged atom. A positive ion has fewer electrons and is therefore not electrically neutral, ie. it is an incomplete atom.

Ionisation	The process of creating ions as in nuclear activity. It can be caused by heat, acceleration or pressure.
Ionosphere	A spherical area in the upper atmosphere, which is highly ionised.
Ionospheric Disturbance	A sudden disturbance of the ionosphere, as from a solar flare.
Local Time	The adjustment of clock time so that the sun is directly overhead at noon. In practice, it operates in time zones, which can give slight differences from universal time.
Magma	A hot liquid type of rock beneath the earth's surface that emerges as lava in a volcanic eruption.
Magnetic Storm	A disturbance of the earth's field, as from the impact of a solar flare.
Mantle	The area of the earth's interior above the outer core and the crust, from 30 to 2900 kilometres beneath the surface.
Mass	The proportionate quantity of matter in a body, which varies with velocity according to the principle of relativity.
Microwave	An electromagnetic wave, between the range of infrared radiation and radio waves.
Neap Tide	A medium high tide when the sun and the moon are ninety-degrees apart at the first and last quarter of the moon phases.

Occultation	When the moon, or a planet hides another celestial body.
Opposition	An alignment of two celestial bodies 180 degrees apart, ie. on opposite sides of the earth
Perigee	When a celestial body, such as the moon, is at its least distance from the earth.
Perihelion	When a planet is at its least distance from the sun.
Radiation	Any type of energy in waves or charged particles.
Retrograde Motion	An apparent motion when a planet appears to be moving in a Backward direction against the star background. It is caused by the relative position and movement of the earth on its own orbit.
Right Ascension	The geocentric position of a celestial body measured in hours and minutes where zero hours is equal to zero degrees Aries in celestial coordinates, measured anti-clockwise.
Solar Flare	A sudden emission of highly charged particles from the surface of the sun, usually from a sunspot.
Solar Storm	A localised magnetic storm on the surface of the sun or in the corona.
Solar Wind	The force of radiation from the sun. It increases as the sunspot cycle progresses.

Spring Tide	A very high ocean tide when the gravitational force of the sun and the moon combine, as at New Moon and Full Moon. The extra low tide is when the sun and the moon are just rising or setting.
Sunspot Cycle	The period of the appearance of sunspots from one period of minimum activity to the next, approximately ten to eleven years.
Sunspot Relative Numbers	The number obtained by a formalised system of counting the sunspots.
Tectonic Plates	The sections of the earth's crust that move slowly over a long period and help to cause seismic activity.
Zenith.	The point where a celestial body is directly overhead.

BIBLIOGRAPHY

Anderson, D.L. *The Plastic Layer of the Earth's Mantle.* Scientific American Offprint. USA (March 1972)

Andriese, P. *Abnormal Animal Behaviour Before Earthquakes.* Open File Report. 80 453. US Geological Survey.USA (1976)

Azimov, I .*Azimov's Guide to Halley's Comet.* Waller & Co. New York. (1985)

Bagby, J.P. *Further Evidence of Tidal Influence on Earthquakes.* The Moon, 6.(Journal) USA (1973)

Ballard, F.M. *Volcanoes of the Earth.* University of Queensland Press, Australia. (1977)

Baranski, S. *Biological Effects of Microwaves.* Hutchinson and Ross, Pennsylvania. USA. (1976)

Barnothy, M.E. *Biological Effects of Magnetic Fields.* Plenium Press, New York (1964)

Bergmann, P.G. *The Riddle of Gravitation.* John Murray, London (1968)

Binkley, S.A. *A Timekeeping Enzyme in the Pineal Gland.* Scientific American Offprint. No 1492 USA (April 1979)

Cape Lyot *Heliograph Results.* Royal Observatory Bulletins, 1V. C 331.UK (1969)

Chandler, T.J. *The Air Around Us.* Aldus Books, London, (1967)

CNSS Worldwide Earthquake Catalog. *Seismic Activity.* Internet (1990-2000).

Davies, P.C.W. *The Search for Gravity Waves*. Cambridge University Press. UK (1980)

Dewey, F. *Plate Tectonics*. Scientific American Offprint. No 900. USA (May 1972)

Dietz, R.S. *Geosynclynes, Mountains and Continent Building*. Scientific American Offprint No 899. USA (1972)

Eiby, G.A. *Earthquakes*. Heinemann Reed. New Zealand. (1989)

Enclopedia Brittanica. *Properties of the Earth–Magnetic storms*. (1999)

Evernden, J.F. *Abnormal Animal Behaviour Before Earthquakes*. Earthquake Information

Bulletins. Vol 76. US Dept.Geological Survey.USA (Nov/Dec 1975)

Eyewitness Encyclopedia, *Space and the Universe,* [CD-ROM] Dorling Kindersley, London (1996) (Right Ascension of the Planets)

Gaskell, T.F. *Physics of the Earth*. Thames and Hudson, London. (1970)

Glasby F, *The Influence of Planetary Bodies on Earthtides and Earthquakes*. Speculations in Science and Technology {Journal) Vol. 2, No 2, p 129—140. Lausanne. (1979)

Gribbin, J. and Plagemann, S. *The Jupiter Effect*. Walker and Co. NewYork. (1974)

Grzimek, B. *Encyclopaedia of Ethology*. Van Nostrand Reinhold. NewYork. (1978)

Gutenberg, B. & Richter, C.F. *Seismicity of the Earth and Associated Phenomena*. Princeton University Press. (1954)

Halley, D.S. Jr. *Earthquakes*. Bob Merril Inc. NewYork. (1974)

Hallam, A. *The Planet Earth*. Rigby, Australia. {1977)

Hooper, J & Scharf, M. *Cosmic Radiation*. Methuen, London. (1978)

H.M.S.O. UK, *Astronomical Almanacs* (!980 –2000) and *Astronomical Ephemeris* (1900-1980).

Hurlburt, A. *The Planet We Live On.* Mc Millan. Australia. (1976)

IPS Radio & Space Services. *Prospects for the Future of Solar Cycle 23.* Australian Space Forecast Centre. (July 1998–2000).

IPS Solar Data. *Flare Reports 1995-2000.* Internet (1999-2000).

IPS Solar Data. *Sunspot Information 1990-2000.* Internet (1999-2000).

Johnston, M. *Normal Electromagnetic Variations, Tectonic Effects, Earthquakes and Animal Behaviour.* Open File Report. 76 149 US Geological Survey USA (1976)

Johnston, M.S. and Mauk, F.J. *Earth Tides and the Triggering of Eruptions* from

Mt Stromboli, Italy. Nature, (Journal) Vol. 239. September (1979)

Laura, R.S. & Ashton, J. F. *Hidden Hazards.* Bantam Books, Sydney. (1991)

Lieber, A.L. *The Lunar Effect.* Anchor Press, NewYork. (1978)

Lithmann,M & Yeomans, D.K. Comet Halley. American Chemical Society,. Washington. (1985)

Lycos. *Volcanoes of the World.* Internet (2000).

Mc Cormack, B.H. & Seiga, T.B. *Solar Terrestrial Influences on Weather and Climate*, D. Reidel Publishing Co. NewYork. (1979)

Mc Kenzie. D.P. & Richter, F. *Convection Currents in the Earth's Mantle.* Scientific American Offprint. No 921 USA (Nov 1976)

Melchiorm P. *The Tides of the Planet Earth.* Permagon Press, N.Y. (1978)

Morth, H.T. & Schlaminger, L. *Planetary Motion, Sunspots and Climate.* Reidel Publishing Co. NewYork. (1979)

Namowitz, S.M. *Earth Science.* American Book Co, USA (1969)

Nautical Almanacs Office, *Planetary Positions and Astronomical Phenomena.* Her Majesty's Nautical Almanacs Office, Royal Greenwich Observatory, London (1900-1980)

NCEDC Data; Earthquakes, Internet. *Earthquake Catalog and Phase Data 1995-2000*. Berkeley University California. (1999–2000)

NCEDC Data; Geomagnetic Data, Internet. *Ionospheric disturbances*. Berkeley University California. (1995–2000)

Nodak. *Current Volcanic Activity 1995-2000*. Internet. (1999 –2000)

Nonevicz, S.M. & Bufe, C. *Atmospheric Electric Field Observation, Animal Behaviour and Earthquakes*. Open File Report 76 876. US Geological Survey. USA (1978)

Oldfield, H, & Coghill, R. *The Dark Side of the Brain*, Element Books, Longmead, UK. (1988)

Ostrander, S. & Schroeder, L. *Psychic Discoveries Behind the Iron Curtain*. Sphere Books, London. (1973)

Playfair, G. & Hill, C. *The Cycles of Heaven*, Plenium Press, London. (1978)

Pressman, A.S. *Electromagnetic Fields and Life*. Plenium Press, N.Y. (1970)

Rauscher, E. & Van Bisc, W.L. *Ambient Electromagnetic Fields as Possible Seismic and*

Volcanic Precursors. Tokyo International Workshop. NCEDC Internet Report. (1993)

Reasenberg,P. *Unusual Animal Behaviour before Earthquakes*. Earthquake Information Bulletin. Vol. 10, No 2. US Geological Survey. USA (March/April 1978)

Rossi, B. *Where do Cosmic Rays Come From?* Scientific American Offprint No 239 USA.(Sept. 1953)

Rossi, B. *Cosmic Rays*. Allen & Unwin, London. (1963)

Sandford, E. *The Planet Mercury*. Faber, London. (1963)

Shave, R.G. *Earth Science*. Reed, Australia, (1974)

Simon, R.B. *Animal Behaviour and Earthquakes.* Earthquake Information Bulletin.. Vol. 76. US Geological Survey. USA (Nov/Dec 1975)

Simon, S. *The Earth.* Collins, London. (1968)

Sky Astronomy. *Planets, Sun & Moon Data,* CD ROM & Internet. (1999)

Smallwood. W.L. *Life Science.* McGraw Hill, London. (1978)

Solar Terrestrial Data. Part B. *Solar Flare Reports.* National Bureau of Standards, Boulder, Colorado.USA (1963/1964)

Tazieff, H. *The Afar Triangle.* Scientific American Offprint. USA (Feb 1970)

Tompkins, P. & Bird, C. *The Secret Life of Plants.* Penguin Books.London. (1974)

Tributsch, H. *Animal Disturbance Before Earthquakes.* Nature (Journal) 606 216 UK(1978)

Tucker, R.H. *Global Geophysics.* English University Press. London. (1970)

US Dept Geological. Survey Reports. *Earthquake Information Bulletins and Phenomena.* USA (1900-1980)

US Dept. Geological Survey. *Epicentre Reports—Monthly Listings.* USA (1900-1980)

US GDES. *Recent Flare Events.* NOAA (1999)

U.S. Naval Observatory. *The American Ephemeris and Nautical Almanac/Astronomicl Ephemris.* Government Publications. (1900+)

Waldmeier, M. *The Epoch of Sunspot Minimum 1976.* Astronomische Mitteilungen der Eidernossoschen Sternwart, (Report) Zurich. No 355 (1977) (Unusual sunspot-minimum phenomena.)

Watson, L. *Supernature.* Hodder and Stoughton, London. (1974)

Wilson, J.G. *Cosmic Rays*. Wykenham Publishers, London (1976)

Wolfendale, A.W. *Cosmic Rays*. Newnes, London. (1963)

Wyllis, P.J. *The Earth's Mantle*. Scientific American Offprint. 916.USA (May 1975)

Chapter 12

Carpenter, G. *Aspects of Astrology*. Beaux Art. New Zealand. (1974)

Carter, C. E. O. *Astrology of Accidents*. Theosophical Publishers, London. (1977)

Collins, P. *The Theory of Celestial Influences*. Stuart and Watkins, London. (1977)

Dewey, E.R. *Cycles-The Mysterious Forces that Trigger Events*. Manor Books, N.Y. (1973)

Fontenrose, J. *The Delphic Oracle*. Berkeley University Press, California. USA (1978)

Gauquelin, M. *The Cosmic Clock*. A.C. Publishers.London. (1982)

Kilner, W.J. *The Human Aura*. Citadel Press. New Jersey, USA (1976)

Leo, A. *The Natal Horoscope*. Fowler, London. (1976)

Mayo, J. *Astrology*. English University Press. London. (1976)

Ostrander, S. & Schroeder, L. *New Discoveries in Astrology*. Prentice Hall, N.J. (1972)

Ostrander, S & Schroeder, L. *Psychic Discoveries Behind the Iron Curtain*. Sphere Books. London (1973)

Ptolemy, C. *Tetrabiblos*. (translated from Greek, by Ashmond, J.M.) Foulsham, London. (1917)

Smith, R.E. *Biorhythm Life Cycles*. Aardvark. New York. (1970)

Tomaschek. R. *Tradition und Forschritt in Klassichen Astrologie.* Eber Verlag. (Report) Germany (1970). (Astrology and biological changes.)

Watson, L. *Body Rhythms.* Harcourt Brace. Jovanaovich. NewYork. (1979)

West, J.A. & Toonder, J.G. *The Case for Astrology.* Penguin. London. (1973)

INDEX

0-595-22641-8